The Pendulum Speaks

It was 1967. I sat in an office in Portland, Oregon. I looked across the large double desk at my co-worker. Sally was six months a bride, a young woman of a marvelously cheerful personality. Her only true ambition, it seemed, was to become a mother as quickly as possible.

Today she had announced the attainment of pregnancy, her round face shiny and her eyes glowing.

"Have you ordered a boy or a girl?" I asked, smiling at the happiness evident in her face.

"My cousin told me how to foretell," said Sally. "She says it never fails."

"And how is that?"

"You tie a piece of string on your wedding ring and hold it over your abdomen. If it swings about in a circle, you will have a girl. If the string goes back and forth, you'll have a boy."

"Let's try it out. There's some thread in the sewing kit."

For a few minutes nothing happened. The ring hung there heavily, like a dead weight.

"Maybe you're pinching the string too hard," I suggested.

Sally loosened her grasp. The ring seemed to pull from her fingers. She tried various holds, twisting the string around her finger, not twisting it, making a loop over her finger; nothing happened.

"Maybe you're supposed to ask," I said rather facetiously. "You have to ask."

"Who?" Sally looked startled. "You know this sort of thing is forbidden by the Church."

"What sort of thing?"

"Spirits. Talking to spirits and occult sorts of things."

"Ask God then," I said. "Maybe He can get a message through."

Sally wouldn't do that. Probably she thought it sacrilegious.

It took considerable encouragement on my part before Sally said, tentatively, "If it be not forbidden, please tell me the sex of my baby."

A wait. Moments went by. A kind of quiver came on the thread. There was a pause, then, slowly, the pendulum began to swing back and forth, harder and swifter.

COVER ARTWORK: Hartmut Jäger
Hartmut Jäger was born in occupied Czechoslovakia of German parents. Near the end of World World II, they returned to Bavaria, West Germany. A goldsmith, jeweller and artist, Hartmut has lived in Germany, South Africa and Australia, where he now resides with his wife and two children. His vivid art reflects his universal outlook and cultural diversity. He hopes that it touches a common cord in all of us.

DESIGN IMPLEMENTATION: Carlene Lynch and Maynard Demmon

Library of Congress Cataloging-in-Publication Data
Kannenberg, Ida M., 1914-
Project Earth: From the ET Perspective / Ida M. Kannenberg.
        p. cm.

                I. Title.

TL789.K353 1995
001.9'42---dc20   95-6504
                                                               CIP

ISBN 0-926524-29-1 : $13.95
                        1. Unidentified flying objects--Sightings and encounters.
                        2. Civilization, Ancient--Extraterrestrial influences.

Printed in the United States of America
Address all inquiries   Wild Flower Press
                                            P. O. Box 726
                                            Newberg, OR 97132
                                            U.S.A.

This book is dedicated to my sister,
Esther Iva Baugh,
who believes in me.

Photo by
Douglas Newman

Ida M. Kannenberg:
•Came from a book-loving family and, like all budding
writers, often spent what should have been food money
for books.
•Has spent a lifetime puzzling over strange and
inexplicable circumstances that life seems determined
to impose.
•Has left a trail of overflow books behind after
seventy-seven years of moving about the western
United States.
•Is now a proud great-grandmother living by the woods
in the outskirts of a small city. She continues to
accumulate more books than the house will hold.
•She still searches always for reasons and purposes behind
the strange and unexplained.

Other Books By Ida M. Kannenberg:
UFOs and the Psychic Factor
The Alien Book of Truth

Table of Contents

Introduction

This story has not been written so much as lived. It is a day-by-day account of my experiences with unknown, unseen personalities claiming to be collaborators with the UFO people—entities who speak to me telepathically at any and every moment.

What they have to say seems to be vital to the welfare and happiness of the people of good planet Earth and the information they have given me has great bearing on the lives and future of us all; therefore, I offer these pages with no apologies for speaking so lengthily of personal experiences.

Some of the material in *Project Earth* has been previously published in my book, *UFOs and the Psychic Factor* (Wild Flower Press, 1992). To omit some of the experiences that I write about in *UFOs* would possibly create confusion for the reader here. What follows in *Project Earth* is the original script, and has been presented in the order in which it occurred.

PART I

The Initiatory Terrors

We both laughed. "Looks like a boy!" (It was.)

In the small cubicle of an office, we two worked alone together day after day with only very brief contacts with the main auditor's office upstairs. Our task was to get out the credit letters on overdue accounts. This we could do with almost no supervision. We had, in three years' time, come to know, like and trust each other.

Once a linen closet for the housekeeper of the hotel, the small office had room only for the two large desks pushed together back to back, so that we faced each other across their steel expanse. A row of file cabinets and an extra chair filled the space behind Sally. In back of me, a shelf held telephone books, discarded from the phone center, that were used to trace skipped accounts. Closed in so tightly together for so long a time, we had few secrets from each other. I knew Sally was the soul of integrity. There was not one deceptive nerve in her body. If she said she had not, by her own will, swung the pendulum of her ring, she had not. Yet it had swung...and rather vigorously at that.

By nature I could not stand a puzzle or riddle I could not unravel. So I made a pendulum of one of my own rings, held it over my wrist, and asked, "Am I a boy or a girl?"

Perhaps having watched Sally's hold on her pendulum I had learned fairly well how not to hold it, for it was only a few seconds before the ring went round and round, ever faster, ever wider in a circle. I could feel it tugging, almost pulling from my grasp. Had I let it go, it would have flown into the wall. "Look!" I cried, "I'm a girl."

We both laughed, laid aside our toys and went back to our typewriters. The monthly statements had to be sent out. There was no more time for nonsense or silly games.

But I began to wonder. At home that night, alone in my small, still apartment, I wondered.

And I made myself a pendulum and two crossed lines on a sheet of paper. The horizontal line I labeled "No," and the vertical line I labeled, "Yes."

And that was the start of it all.

I had recently left my husband because of sheer incompatibility, believing two persons with two completely different senses of value should never marry in the first place. They can differ in any other way—size, shape, age, nationality—but if their values are different, the whole story is finished before it is begun.

I was not at all heart broken over the dissolved marriage, for the quarrels and disputes had finished us long before. I was disappointed in myself, however, because I could not hold things together and somehow overcome our differences. I felt that if maybe I had been wiser or stronger or more courageous, I could have overcome any difficulties. In the end I gave up and moved out of his house and life, and I was now engaged in making a life for myself alone. I did find the long evenings lonely, however, for I was not a socially busy person.

I preferred to read, sew, study or write. Therefore, I had many spare evening hours to work at my pendulum "experiments." In the beginning, I could only ask the usual groping questions, thinking that no matter who or what was swinging the pendulum they could certainly answer them.

"Am I going to get the raise I want?"

"Yes," the pendulum indicated. (I did not get the raise.)

Always I would start out by asking, somewhat unnecessarily, "Is someone there?"

Invariably, "Yes," was the answer. Of course, there was someone there or I would not get an answer, I thought. I felt stupid for having asked the question. But it did seem a good opening, so I used it each time.

"But, of course, it is only my subconscious," I said to myself, "Though I'm not quite sure what my subconscious might be, they tell me I've got one." Quickly I figured it all out. "It is my subconscious controlling my muscles, which respond ever so subtly by swinging the pendulum." Somehow that did not seem to resolve much either, but for the moment I left it at that.

Most of all I wanted to ask, "Who are you?" But an answer of yes or no could not reveal that. I had to content myself with such questions that could be answered "yes" or "no." This I found limiting and most unsatisfactory.

After stumbling around awhile and getting very misleading and disappointing "yes" and "no" information, I became disgusted and dropped the whole thing temporarily.

At this time, my personal life came into a state of complete upheaval. Already there had been the marital separation; now came divorce. I started a little antique shop, which I could keep open only for a couple of hours in the evening and on Saturday. Results were encouraging, and after a few months I gave up the office job to keep the shop open full-time. Then I met Bill, a darling man. It took all my time and attention to get the shop running efficiently and to have some time with Bill.

The idea of the pendulum was pushed into the back of my mind along with something forgotten from my childhood. It was a memory of the time my father had brought home a Ouija board, back in 1921. Mother, Father, Esther (16), George (11), and I (7) sat around the big square oak table in the kitchen. I watched in anxious anticipation as Father removed the board from its box and explained its use.

The table was covered with a bright, red-and-white plaid oilcloth. There was a kerosene lamp in the middle. This was the center of family life in the evenings. Father read the paper and Mother mended or sewed, while we three children did our homework or read.

Now we pulled our chairs a little tighter, and everyone took a turn at the board. Nothing happened. We asked questions, read directions, laughed, joked, ate popcorn and tried again. Mother barely touched it, then refused. "I don't think it's right," she said. "There is something strange about it."

Father read the directions again.

"Here," he said, "it says if one alone has no success, two should try it." Since Mother refused to touch it again, Esther, as the eldest, was elected. She and Father put their fingers to the planchette and asked, "Where are the lost dominoes?"

Nothing happened.

Their fingers pushed down harder. "Not so hard," said George. "You're stopping it." His freckled face was alight with wonder and curiosity.

Then the planchette seemed to lurch and skid. It jerkily spelled out, "Closet!"

"Father, you're doing that!" cried Esther.

Father looked a trifle pale. "No, I'm not. You are," he said.

"But I'm not!" cried Esther.

There ensued a heated argument. Each thought the other was teasing and pushing the planchette. Father settled matters by picking up the gadget and putting it back into its container. It was never used again.

"Why don't we look in the closet?" I asked, believing that to be the only logical thing to do. Since I was only seven, no one ever paid much attention to my suggestions. Later, I furtively looked in the closet, but the dominoes were not there. We never did find them.

Several times I asked my parents if I could play with the Ouija board. I was sure I could make it work if they would only give me a chance, alone and by myself. There always seemed to be some reason why it was not convenient to take it from the cupboard. I began to believe Mother was thoroughly scared of it. (She was very intuitive.) After a while, I stored the Ouija board in the back of my mind and "forgot" all about it. Nothing in the nature of a personal experience with psychic phenomena crossed my path again for the next 46 years.

Then came my experiments with the pendulum and the dissatisfaction with the "yes" and "no" answers.

Somewhere in the back of my mind, the Ouija board and the pendulum must have gotten together, for the next time I thought of the pendulum, along came the idea, "Why not make a drawing of the alphabet, like the one on the Ouija board? Then the pendulum can spell out its messages!"

It was an inspired thought. (Later I came to know just how it had been inspired.) Right then, I drew a circle around a saucer. I drew the horizontal "no" and the vertical "yes" lines, and, starting at the nine o'clock point, drew the 26 letters of the alphabet all the way around the circle.

At first, the pendulum was balky. It started, stopped, stuttered, and gave ridiculous, patently false or misleading answers, refused to answer at all, or demanded answers of me!

"Who are you?" I asked it. At first I spoke all questions aloud. Later I learned to whisper, and then only to think my questions and the answers to their questions.

Such concentration it took! And that was the opening door, the secret to the whole process. Much later, I came to realize that by concentrating to the oblivion of all else, I was hypnotizing myself and opening the door to telepathic communication. I discovered that hypnotism is simply opening one's mind to another mind in order to be impressed by it in one manner or another (not controlled). In this case, the impression was telepathic. But telepathic communication with whom? And why? And how?

Now even more changes came into my life. Mother was widowed for the second time, and she wanted me to move with her to a smaller town 100 miles away. We found a nice apartment there, and I opened another antique shop. I had received a little money and was able to buy a better inventory, which meant more business.

My friend Bill made frequent trips to see me and called daily, sometimes several times a day. (His phone bill must have been horrendous.) But I was still lonely. I cast about for something to take up my spare time during slow hours in the shop.

I told no one about my experiments with the pendulum. I conducted them in the privacy of my bedroom at night, after Mother, an early sleeper, had retired for the night. Certainly Mother would not approve of such doings! One night I asked, "Can you locate my Father?"

"He is still asleep," was the astonishing reply. *"Wait a little."*

After a very brief time, the pendulum continued, *"Ida!"*

"Who are you?" I whispered back.

And the pendulum swung back "E. J. G."—Father's initials.

I contemplated this very seriously. I did not want to be taken in too easily.

"What is your first name?"

"E-R-N-E-S-T." (That was correct.)

"Your second?"

"J-O-H-N."

"And your last?"

"G-R-E-E-N."

Certainly correct! It made one think! Just what was going on here?

I tried to think of something only Father would know. I asked several questions. The pendulum answered them all correctly. I still was not satisfied.

"If it is my subconscious," I thought, "of course it would give me all the right answers, since I know them."

"Can you tell me the names of some of the books you used to read?" I was thinking of a series of books that he had owned and read over and over. But, surprisingly, a title came back that I did not recognize.

"Zorn," said the pendulum.

"I don't know that one. Was it by Zane Grey?"

The pendulum started something, stopped, then went off into something else entirely. I made an angry retort.

The pendulum indicated, *"You are an impudent brat!"*

I burst into tears and cried, "Now I know you are my father!"

I began to feel the pressure of time. I was eager to ask a thousand questions. I could not choose one.

"Shall I tell Mother about this?"

"No," swung the pendulum madly. "NO!"

"I want to see you," I begged, thinking that the spirits of many deceased persons had been reported to visit the living.

"You are too young to die," was the amazing response.

And that was the last contact I had with that personality, father or not.

Several times the pendulum answered my question, "Who are you?" with the initials E. J. G., but when I asked questions, the answers were not satisfactory—the feeling was just not the same, not right. If he had been there at first, I decided, they were only pretending afterwards to keep me interested.

I tried to question my mysterious pendulum on the status and place of being of such deceased loved ones, and I even asked questions about Jesus, but the response was a most curt, *"Do not ask about divine matters. Ask anything else."*

Another time it sneered (about something quite different), *"You are too stupid."* I cannot tell how I sensed the derision and the scorn that accompanied the spelling out of these words, but I could definitely feel contempt.

On more than one occasion, I teased it. When I did, it indicated succinctly, *"Go to hell!"* And there was real anger in the force of the words. Of course this did not make the slightest bit of sense—but then none of it did.

Sometimes in the middle of spelling out, the pendulum would stop abruptly and hang motionless, as though waiting for some event to pass. It seemed to be very heavy then—much too weighty for so small a thing, and the string felt as though it would pull out of my fingers.

Unfortunately there were many interruptions at the shop. Customers came into the store, the phone rang. I would have to stop in the midst of something quite interesting. But none of the responses ever had much importance. They never got anywhere. Many times I said, "This is just plain stupid and absurd. I am not going to waste one more minute on it." But then I would wonder, "But what is going on here? How does this work? Who or what talks to me?"

Sometimes, in the night, I would awaken to feel the bed "move" or my body would be vibrating strangely—not actively shivering or shaking, but vibrating inside, as though all the molecules were being given a good shaking up. But how, why, by whom and for what purpose? I could determine no real sense in any of it.

Almost from the first, a strange sensation accompanied the use of the pendulum. This would usually happen in the evenings as I worked alone in my room, ostensibly reading, but actually playing with the pendulum. Every so often, I would hear a thud or a series of thuds in the air near the floor. I thought someone powerful and important was coming on the scene, with "a tread like thunder on the air." I thought all the way from God to Satan as to whom it might be (but because of various other happenings, I knew it could not be God.)

Eventually I became panic-stricken. It might be Old Nick himself. Sometimes he seemed to be accompanied by pernicious little nippy things that I sensed to be like super huge gnats, or even human-looking gnats! I called them the "Flits."

"The Flits are back," I would say when this phenomenon appeared. Later, all this disappeared when other elements took over. I knew very well I was not actually seeing, hearing or feeling any physical manifestation; it was a mental sensory experience (if that description makes any sense). Nothing else did.

Communication through the pendulum became increasingly wordy. Still, it was slow and tedious to try to get a sustained meaning. Messages were often interrupted, and the words went swinging off in a new direction. There was somehow a cross communication, as though two factions vied with each other for my attention.

Once, quite early on, I called one of my communicants by someone else's name and he quickly corrected me—bluntly too.

"I'm sorry," I said. "But how am I to know which one of you is speaking to me if you do not announce yourselves?" There was a very still silence for a little, then the pendulum asked,

"You mean you cannot tell who is talking?"

"How can I tell?" I asked sharply. "All I can see is a piece of metal and a string moving to certain letters. What is there to tell me who is on the other end of the line?"

A rather long silence followed this and I wondered, "What did that mean? Why did they think I could always tell them apart? And what difference does it make if I can or cannot tell who is speaking?"

Something about that last question began to nag and worry me. Aren't they all working together from the same place? What does this mean?

Sometimes, in the middle of the night, I would be awakened from a sound sleep by the greatest impulse to go to the pendulum. By this time, one of my correspondents had named himself Hweig[1]. Hweig seemed to be in control of these late sessions, and he began to lead them off into paths of rather intimate questioning.

"Are you married?" he asked surprisingly one night.

"Good heavens!" I exclaimed, startled. "As I told you before, I have been, but I am not now."

By now I was thinking of the pendulum no longer as merely a piece of string, but as definite persons or personalities. I was only befuddled about who they might be and where they were from. I asked endless

---

1. Pronounced: Hw-eye-jsh.

questions, all of which were imperiously laid aside while they attacked me with question after question of their own.

Hweig and his invasion into my private life was the most dominant at this period.

"Do you have a lover?" he asked.

"That's none of your damn business!" My answer was spontaneous and angry. Then, wanting to be cooperative, and partly out of curiosity and partly because I did not want to lose this strange contact, whatever it might be, I responded, "There *is* a man I am interested in, and he in me and that is all there is to know."

I felt it advisable to answer their questions as truthfully as I could, not only because I am of a truthful nature, but because they must have a true picture of me if our communication was to be valid.

"Then I will be your lover," said Hweig flatly.

For the first time a chill of cold fear shot through me. All the things I had ever read about demons and possession swept through my mind like an icy wind.

"I don't know who or what you are," I said, and my voice quivered. "And I don't want some kind of spirit lover. Let's not talk along these lines."

"It's too late," Hweig said. *"You said I could enter your mind and understanding, remember?"*

I was growing frozen and rigid with fear. "Something like that, but this is not at all what I meant!"

"You said enter. Now I am here. It is too late to do anything about it." The pendulum swung with terrible power.

"Now wait a minute! I don't have to sit here playing with this stupid string. Good-by!"

"Too late," began the pendulum, but I flung it down, shouting, "Go to hell!"

Many times in the future I was to throw it down again and again, shouting, "Go to hell!" but each time some terrible impulse made me go back and pick it up again, usually saying, "I'm sorry, but don't talk like that. It spoils everything."

Each time the session resumed on more staid matters without apology or acknowledgment of mine, but each time, sooner or later, the subject came around again.

"You're mine now. I am with you always."

"That's invasion of privacy. You can't do that to people."

Tauntingly, *"There is nothing you can do about it. You said enter. I am here."*

Then, abruptly, this tormenting ceased. Only occasionally a little mocking statement, like an echo, repeat, *"You are mine now."* And, eventually, due to the daily pressures of business, home life and visits from Bill, I nearly forgot about it.

There were developments from the other personalities "on the string." Every day I had a fierce impulse to go back to the pendulum, and every day there was some new surprise.

Then they began to give me instructions in my personal life, even saying, bluntly, *"Go take a bath."* At first I complied every time, still calling it curiosity, experiment—still trying to find out who they were and what they really wanted of me.

One Sunday afternoon when Mother had gone visiting and I was alone at home, my invisible instructors told me five separate times to take a bath. At first I complied, but the last time I protested.

"This is just too ridiculous. I'm trying to be a good sport and go along with this thing as much as possible, but five baths in one day, even quickies, is completely and absolutely absurd. Go take a flying leap in the river yourself!"

The pendulum stopped dead. There was a sudden hiatus, a blank silence. Nothing moved. Then, almost at a creeping pace, the pendulum spelled out, *"Do you hate us that much?"* It was almost wistful.

What in the world! It had only been a slang expression—certainly nothing to be so emotional about.

"I'm sorry," I said, feeling very contrite. "I should not be so rude. I didn't mean anything by it."

By this time I was reading all the books I could find dealing with psychic phenomena. This was years ago. Not so much literature was available as has been published recently. There was Edgar Cayce, Ruth Montgomery, Robert Ford—all the old standbys. I also dipped into many related and fringe related books—the Egyptian Book of the Dead, the Vedas and all of the contemporary magazine writing I could find.

Then another note came into the experience. *"We are preparing you,"* they said. They did not say for what. Sometimes they said, *"Take off your glasses, and stare at the wall."*

I would focus my eyes intently for some minutes, but soon the feeling of urgency would leave and I would more or less forget what I was doing, put on my glasses and do something else.

At times, when I was in the store, they would say, *"Go to the window and look out."*

Then I would have a prickling sensation on my skin, as though the sun were shining on my arms and face. Presently I would forget what I was doing and wander away, back to my desk or to dust or arrange shelves. Often a customer would come into the store and we would become involved in the merits of a piece of custard glass or Chelsea china.

The shop had very large plate glass windows and one door. Sometimes, when I was sitting at my desk using the pendulum, the big windows would crack or pop, quite loudly.

"Angels are coming through to watch you," said the pendulum.

I could certainly detect the falsity of this. It was even more detectable when they insisted, *"Now there are archangels coming through. There are about forty of them in here now."*

This sort of thing disgusted me. I really could not decide why I kept on with it. They were so uninformative, so evasive, so untruthful, so imperious, so downright nasty at times. No matter how many times I threw down the string, shouting, "That's it! I've had enough!" I always came back for more. I was impelled. I was not able to say "No" and stick to it. Like a child with a new and fascinating toy, I played with the pendulum constantly, every chance I got—even to having it under my pillow at night.

Never had I known such a compelling force, keeping me at a task I did not understand. There was soon no let up from these personalities, night or day. Every moment was, in one way or another, given over to direct communication, or to trying to carry out their demands, however absurd they might seem. I felt that I was trying to enforce some unseen, unknown law of development. I was being developed for something—*"prepared,"* they said. But for what and by whom?

Every so often, Hweig would break in with some kind of intimate conversation, but I began to detect something else—another personality that was beginning to make itself known, but seemed not to be detected by the others. And I sensed that this new personality (who never had a name) was somehow protecting me, or trying to. It would break in on Hweig's words, send his communication into a spin so it was never completed. There seemed to be now at least three separate factions contending: Hweig, the other early ones and the new protective one that was never named or openly acknowledged. Somehow I "just knew" I was not to give away his presence by addressing him directly or referring to him.... He was just there, like a guardian angel.

Some of the other personalities began to clamor for more attention. One kept repeating, *"Now you are going to be initiated. You must repeat what I tell you three times. This is called the Rule of Three. You must vow to tell no one of what we say. Vow! 'I will tell no one!' Repeat it three times, then say 'Amen' three times."*

I had to consider that. It seemed harmless. What could "they" possibly tell me that would be harmful if I were to keep it secret? After long thought, during which I paced round and round the shop, I finally came back to my desk and said, "All right. I vow. I will tell no one. I will tell no one. I will tell no one. Amen, amen, amen."

I must have taken that vow very seriously, for afterward I could never remember one word of what I had vowed. Never could I recall what the secret was—not one inkling. There were many such secrets and vows, but never afterward had I the slightest clue as to what I had sworn not to reveal. It had to be "good" things, however, for each time they told me to vow, I asked, "Is it evil? I will do no evil."

Each time they said, *"It is good."*

"Is it against the will of God? I will do nothing against the will of God."

"No," swore the pendulum. *"It is not against the will of God."*

"Then I will swear. I will tell no one." And once again I would repeat the formula for secrecy. And the memory of those vows, their contents, are forever sealed, even from my own memory.

One of the more ludicrous series of persuasions was that of my spiritual status. After an interval of questioning and many vows to "tell no one," it was announced that I was thereby a saint! After another period of time and some equal rigmarole, I had advanced to an angel! Still later, I became an archangel, and lo and behold the time came when I was a *goddess!* My response to all this was laughter. Even in my confused state, I knew all this was fabricated malarkey. Perhaps this angered my unseen promoters, for after a time it was announced that I was no longer a goddess but merely an archangel, then came demotion to an angel, and finally to a mere saint. My elevation had been brief, but I was still able to laugh and take everything as a kind of game—a hilarious pastime. But I wondered what kind of correspondents I could possibly be dealing with. Were they from my own subconscious? Were they subsidiary personalities of myself? Mischievous spirits? I could not guess, and I did not have enough knowledge to be afraid.

Every time I made a vow to keep a secret, the pendulum would say, *"Ask for something."* I thought and thought, but could not come up with anything that seemed appropriate. A few, rather facetious, requests met with very complicated replies. I was not sure if I was being offered something in return for the vows of secrecy I had made, and I certainly did not intend to be "bought off" by something or someone I knew nothing about. Finally I said, "Whatever you have to give." Strangely I had the sensation that the pendulum, or the force behind it, was delighted with this. But I don't know what the "whatever" was I was supposed to have received. A secret gift? A secret, certainly!

By this time, I had evidence to reassure me that this could not be my subconscious alone. Impossible! It either had to be spirits of the dead or some kind of hidden personalities.

Could it be parts of my own consciousness, split personality, dissociated secondary personalities? I read several available books on this sort of thing and it just didn't add up. This experience was totally different from anything I read.

The only other source I could think of, other than from the spirit world, was that it might be personalities from another plane of being, a world counter to our own. Some of the books I read suggested this possibility. One in particular discussed a world coexistent to our own, occupying the same space but existing in a different time dimension. I pondered this, and decided it must be something close to this explanation, but I could not quite grasp it. I could not understand. Perhaps my awareness had not reached the point of development that would let it make sense to me. Always the tenacious skeptic, I demanded that things "make sense." Coexistent worlds did not.

I fell back on spirits then. My communicators did not help much. Once, previously, they had said, *"We are poor lost souls in hell."* But I had detected something not only spurious, but teasing in an unfriendly way.

Now they said, *"We are a group of men here in your city that have learned to do this."*

That sounded halfway plausible, for some reason; at least it made more of the sense I was demanding.

Now my living arrangements changed again. Mother remarried. The ceremony was at my daughter's house and it was lovely. Several times during the evening, I saw my new stepfather watching me closely, staring.

I wondered if I might be acting strangely, so that people were beginning to notice. I went into the bathroom and studied myself in the mirror. I decided I looked my very ordinary self—brown hair with a few threads of gray, brown eyes (hazel in certain lights), high forehead, glasses, snub nose, cleft chin, tall thin body. Not bad for my fifty-six years. Many times in my life I had been asked if I were a school teacher, so I suppose I looked like what people think a school teacher should look like.

I did look a little strained now, and even more underweight than usual. I had not been eating much. Mother had been complaining about that. But I could not see that I looked in any way nervously overwrought or tense. Nevertheless, my stepfather watched me all evening and several times he made it a point to draw near and ask how I was feeling. Perhaps Mother had noticed me acting withdrawn or strained and mentioned it to him. Mother had said nothing to me except to complain about my lack of interest in her good meals.

A new intensity in the relationship was about to begin. Had I known then what I was soon to learn, I would have buried the pendulum in the deepest pit.

On the Edge & Looking Over

I rented a little gray house in an alley only a block from the shop, and had all the long evenings alone to communicate with the invisible personalities. One evening, the pendulum told me to bare my back and sit quietly. I thought I could feel something going on in the muscles or nerves in my back. One shoulder relaxed, then the other. Cords were very lightly manipulated—nerves or muscles or something, I could not ascertain what. My back had given me much trouble due to an old injury. Once, very mysteriously, I thought I heard someone say, inside my head, *"Poor girl, that is the seventeenth time they have done that to-day."* What in heaven's name could that mean? How could I hear something like that? How had I heard it? Certainly not with my ears! I could not imagine, but my thoughts went round and round.

Once, while I was reading in bed, I looked down and thought I caught a glimpse of a heavy thread or cord being drawn across my hand. It was white and, surprisingly, square! I could feel it being pulled across, but could see nothing that could be doing it. There was a chill, moist sensation where it touched.

Evenings became a pattern of intense effort to find answers to my many questions (most of which were totally ignored) and at the same time to try to carry out feats and experiments they asked of me.

Sometimes they would answer one of my questions, but the answer would be so outlandish that I could only too quickly refute or disprove it. When I angrily denounced the fraud, they would say, *"That was a test."* But I never knew what they were testing for.

Once, in the shop, the personality using the pendulum said, *"Do you see that tool on the counter behind you?"*

I looked about, but could not determine what they meant. I asked them to describe it more fully.

"It has two long points and a handle."

"Oh, the carving fork."

"Plunge it into your chest," they ordered.

"I will not! Do you think I'm crazy?" I shouted vehemently.

"It is a test," they said. *"Forget it."*

The slow swings, the errors of trying to follow the exact line of the pendulum, the interruptions, the general tediousness of the communications made me ask, "Is there no other way we can communicate?"

"No," said the pendulum flatly.

Soon the demands on my attention increased. Somehow impelled to return to the pendulum ("the work," they called it) again and again, I had time for little else. I was head over heels involved in just that, and the rest of my life's responsibilities and duties and my health suffered from neglect. I was sleeping less and less and was scarcely remembering to eat, now that Mother had gone to her new home and was not hovering over me.

Once, as I sat on the edge of my bed concentrating on the pendulum, I heard a swish in the air, as though someone had opened a heavy door into a room filled with a different and busy atmosphere. A man's deep voice said, *"Do you remember the little girl under the lilac tree?"*

Another said, *"Will you drop this off for me on your way down in the morning?"*

And a third, very masculine, fine, deep voice said, *"Wait. She is beginning to hear us now."* And then silence, as though the door had been swiftly shut.

Now subtleties and nuances of expression began to confirm for me that there were various personalities involved. One, who gave himself no name invariably began his messages with, *"To be frank..."* So I began teasingly to call him Frank. "Oh, it's Frank again!" This somehow seemed to disconcert him, for his replies would break up into choppy phrases. Maybe he was laughing—at least that was my impression sometimes. You can see how deeply I was becoming involved in these messages. I was certain I could detect emotional responses more than were exhibited in the words themselves.

Another frequent communicant always ended his messages with, *"Be of good cheer!"* Often he broke into others' sentences when I became upset by something that was being said. Like oil on troubled waters, his messages came swinging through blithely, breaking into and carrying on as though he had more power of some kind, more force than his comrades. I began calling him the Good Cheer Man and he did not seem to mind. I could almost detect a chuckle hanging in the atmosphere.

There was another, though, who gave himself no name, but was extremely obnoxious. He did not like it at all when I said, "Now listen, Buster! Don't get smart." There would be a second's hesitation in the pendulum, then it would come swinging back in a puzzled tone, "Who is Buster?"

This made me burst out laughing, I guess because I thought I had out foxed them at last. Now I was making them guess. But presently, since it did seem to worry him, I confessed that Buster was just a name we used for someone who was making things a bit difficult. After that he was no longer quizzical when I used it, but rather angry.

"Don't call me that!" he would demand.

"O.K., Buster!" I would laugh and drop the string so he could not express any reaction. He seemed to express considerable derision and scorn and seemed to actively dislike me. He made nasty little remarks about my general intelligence and was not at all flattering. Since I had passed intelligence tests that permitted me to become a member of Mensa, I felt I could not be the most stupid person on earth (though I thought intelligence tests in general were rather dumb themselves).

The Good Cheer Man, on the other hand, was unfailingly polite and well mannered. He never flattered me, but tried to smooth over the hurts and stings given by "Buster" with some kind of compliment I felt to be sincere.

"Do not be alarmed," he would often say. Or, *"Have no fear. You cannot be harmed."*

I was increasingly not so sure about that. There were still messages from Hweig that did alarm me in their implications. *"You cannot escape me,"* he seemed almost to whisper, the pendulum swung so lightly. *"You are mine."*

I began to protest, loudly and often. "I will not be taken over by another personality. I will not have my mind invaded. I will not become a robot to your usages. I will not be a puppet while you pull the strings!" Loud and vigorous were my declarations, but with a growing terror in my heart.

"You cannot escape now. You invited me to enter." Hweig's whisper declared.

Then the Good Cheer Man, swift and strong, said, *"Have no fear!"*

Which one was I to believe? Was I some kind of a battlefield for contending spirits? I felt the only thing that could combat this terrible growing conflict in my mind and heart was to discover who, what, where, why.

I demanded. I wept. I begged. Then I grew strong and demanded again. "I can quit this any time, you know!" Only I could not stay away from the pendulum. I always came back.

They gave me an address in town. *"Come to this address. We are gathered at this house and you can consult with us face to face."*

I verified the address with them over and over again to make sure I had taken it down correctly. I called a taxi, locked the store, and left a note on the door, "Back in one half hour." I gave the driver the address and sat back, heart pounding.

For one hour we searched. The address was of a vacant lot. We tried all kinds of similar addresses, even though I had verified the address before leaving the shop. I still intended to give it every possibility of being real. Actually I had never felt it was real, but I was in every way willing to try to verify it. Finally we gave up and went back to the shop. I explained to the driver that someone had phoned me with antique items to sell and had wanted me to come at once to see them. "Someone's idea of a joke," I said bitterly.

When I chided my communicants for the false lead, they replied blithely, *"It was a test."*

"If you are testing my gullibility, you have your answer," I said. "If you are testing my willingness to play along with you, you have about tested it to its full strength. My willingness is about at an end. So is my patience with you."

"Be of good cheer," said the pendulum. *"It is all part of the work we must do."*

"Well, I don't have to work for a bunch of inconsiderate idiots who don't have the slightest idea of fair play," I said. "I am just plain fed up."

I neglected the pendulum for a while, but try as I might, I eventually came back and the whole routine took up just where it had left off, as though I had never been gone. Once, I remarked on the passages of time, when I had to be gone a long while in the middle of a good conversation. They responded, *"Our time is not the same as your time. When it seems like hours to you, it can seem like a few seconds to us."* They refused to elucidate further right then, but much later they explained time in more careful detail.

At last they had said something sober and sensible, as though I might hope for something serious and purposeful to come out of all this in the long run. It was most encouraging.

An alternate to the pendulum when I was not near one was a kind of scratching I could do. I would conceal my hand beneath a table or paper and scratch with my fingernail on whatever was there. In the vibrations of the scratch, I could detect definite words.

Later (and this was the most absurd manifestation of all), I could click my teeth together and hear words transmitted to me in that vibration. At first it was fun. "I have the only teeth wired for sound in town," I thought. It seemed there had to be some kind of vibration present for me to receive words. Quite a lot of hocus-pocus went on about the teeth until I became tired and tried to stop it. I could not. The teeth clicked now without my will. This frightened me, so I seized a nail file lying by, and filed at my front teeth. The clicking stopped, as though they realized I was serious and might do more damage.

After some thought, I decided I had to have some touchstone, some element to keep me anchored in my own being, time, and mind. So I bought a Bible. I was not very religious, but I did believe in God and in His help to those of His creatures who tried to obey His will. The book gave me the greatest comfort—not for anything I might read in it, but just for something at hand to give me a sense of security.

Usually I ate at home, quick snacks that could hardly be called a meal. But sometimes the loneliness was too much even for a natural-born loner like me, so I went to a nearby restaurant where I could at least look at other people while I ate. I did not want to talk (I avoided my family), but I wanted to be around others, so I selected a favorite booth for myself at the back of the restaurant. There was some sort of cooling device for juice or drinks that ran rather noisily near my booth, and the vibrations from it seemed to somehow increase those vibrations I felt inside my body. These had become a permanent and constant fixture, but were low and subdued until I came within range of this vibrating cooling machine. Then the response in my insides was terrific. A vibration was set up that threatened to tear my body apart. Not just my head, but my back, stomach, abdomen, arms, legs—all of me. I felt I was being torn to pieces by this vibration. It was too much to bear. I went back to those sketchy meals at home that were only a bite—a cup of coffee and a cheese sandwich. I just forgot to eat more.

All this time I was in daily contact with Bill. Some weekends he came to see me, sometimes I went to see him. On one of these Sundays, I told him about the pendulum. I didn't tell him much about my experiences, just that I was receiving messages from some unknown source. He was the world's most dyed-in-the-wool skeptic; he believed only what he could see and touch himself. So we made up an alphabet and a pendulum and I showed him how it worked for me.

"You are doing that," he said. "I see your muscles jerk."

"Sometimes there is a quick release of muscle tension," I answered. "But I am not making the pendulum swing."

"It is your subconscious," he said. "You don't know you are doing it, but it is involuntary muscular action, spurred by your subconscious."

We argued. He tried to use the pendulum but it stopped dead and did nothing.

"You see," he said, "it doesn't do a thing for me. Not a twitch."

"Because your mind is shut," I argued. "You've got to have an open mind. You've got to believe it is possible, or at least that it is not impossible."

"Belief is bunk," he said shortly. "Belief is what has ruined the world. Facts! That is the only thing I know. Facts!"

"The fact is it works for me," I said.

"The fact is your subconscious works for you."

We argued rather angrily. I could feel the inside of my head getting more and more tense. There was a hard lump in it, as hard as a football. I tried every way I could think of to prove what I was saying, or at least to make him open his mind. He had to believe! But he was adamant. He could not make the pendulum work; therefore, it was all phony (including me).

The argument grew more tense, as did the inside of my head. Then, without knowing what I was going to do, I threw myself on the floor as though in a faint. I was fully conscious, but I could not move or open my eyes. With trembling hands and cold water, Bill tried to revive me. I began to cry. He carried me upstairs and laid me on the bed. I continued to sob for some time, but I could not bring myself to tell him the many terrifying problems I was experiencing. By that time they had become terrifying. Perhaps my vows of silence had something to do with that.

"Ida, what is wrong? What is going on here?"

All I could sob was, "Help me. Help me. No one else can. Only you can help me."

He held me for a long time. Of course he could not help me. He had no way of knowing what was wrong, and I could not tell him. I went home that evening quiet and frightened. This whole thing was taking on a terrifying aspect. It had gotten completely out of my control. For the first time, I realized I had no control. It had overpowered me.

About this time still another factor entered the picture.

Every night, after I closed the store, I hurried home to my bit of supper and then either to read or to work with the pendulum. I was so intensely curious about whom I might be talking to. What could they tell me about worlds beyond my vision? Would they ever reveal to me the true condition of their being? All I could imagine them to be was spirits of the dead, but there was too much about them that belied this explanation. Could they possibly be inhabitants of another world? Where was the world then? How did they move the pendulum? By what laws and energies did it work for them, and for me? Why were they so inconsiderate of my well-being, of my state of mind, of my happiness? If they expected so much response—"work" they called it—why could they not be a little more careful of my health and peace of mind? What was I to them, after all? A guinea pig? A case for experiment only? Didn't my feelings count? Who were they to assume such control over my life, practically my every movement, and yet treat me with no respect as a person—to treat me with scorn and derision, sometimes even to the point of offering obscenities? (Although the Good Cheer Man usually broke in and scattered comments of the unwelcomed kind, he was the kindest, most considerate of the lot. Even though he seemed not to understand my way of being too well.) Sometimes it seemed that they did not really know me any better than I knew them. This was baffling, indeed.

I had lost control over my actions to the extent that I could not stay away from the pendulum at any time. It encroached deeper and deeper into my nights. I sat up until two or three in the morning, asking questions and receiving rebuffs, demands and instructions. I found myself increasingly unable to disobey. Not that they ever asked me to do anything that would harm anyone, and really there was no harm to myself, except for the increasing lack of food and sleep.

Now there were many times I could detect subtle changes in tone behind the words of the pendulum. How could this possibly be? I could not imagine. How could tones be coming from the swinging of a small silver cross and a piece of sewing thread? (I thought the cross would be protection against evil influences.)

Then, all at once, I realized what was happening. I was not just watching the pendulum, I was somehow hearing the words in my mind! It had developed so gradually I had not realized it until there it was, full blown—a kind of telepathy from somewhere. But, oh my God, from where? Who were these forces or personalities who spoke to me from beyond the physical shell of this world? What were they up to?

It was so hard for me to believe in the "mind hearing." I would hear a sentence or thought clearly in my mind, but I could not accept it as a fact. I kept on using the pendulum, but it only tediously spelled out what I had already heard in my mind. There were no physical vibrations of sound, just a quiver of meaning.

One day, the personalities asked, *"Do you see that fine white house on the corner?"*

"Which house?"

Presently that was determined. *"Buy it."*

"Buy it! With what?"

"The money will come."

"Well," I thought, "here is a practical opportunity to test my communicators, and to see if there is anything at all in what they say."

So I went to the house and asked the tenants who owned it. It was owned by a real estate broker, located just around the corner. I went to ask him if the house was for sale.

"Are you some kind of a psychic, lady?" he asked. "I just decided this afternoon to put it on the market."

We talked terms and I went to the bank. I guess if I had not been too timid and insecure to try it, I could have swung the deal. It would have meant spending my last dime on a down payment and closing costs. I did not have enough confidence to do it, either in myself or in the promises of my communicators. It would have been an excellent investment, but no money fell into my lap that would have paid for it. Perhaps if I had purchased it, if I had trusted them a hundred percent, the money would have come. I could not, however, give myself over to them so completely, and I decided against buying the house.

Another time, I was directed to a certain bank, where, amid the bushes along one side, I was to find something important. Directions of where to go came into my mind as I walked along the street. *"Turn here, look there,"* they commanded. I believed it vital to check out each bit of information, no matter how odd it might sound and even though I placed little credence in its veracity. It was a clue to the source of the information I was really looking for.

One morning, as I prepared to go to work, the voices told me to find a quiet place. After some discussion, I decided the Catholic church was the most quiet place I could find. I passed up the shop and went on to the church, several blocks away. I made the sign of the cross and knelt in the pew a few minutes, but could not think of a prayer. The inner voices kept telling me someone would come to meet me there. I sat in the pew, fixed my eyes on one place on the altar and waited. I found myself absolutely motionless. I did not move my head or my eyes, but remained staring at one spot. As time wore on, I fretted a little, but never moved. The voices kept reassuring me someone would come, and that it was very important for me to remain quiet and wait. Once a man came in and sat not far from me. I thought surely this must be the one I was waiting for. He sat for quite a while, but then left. I soon became restless again.

Finally I announced, inwardly, that I was certain nothing was going to happen and I was leaving. I did not meet any opposition to this.

All was silent. So I looked at my watch for the first time since I had come in. It was after three o'clock! I could not believe it. I had sat motionless, breathing most shallowly and eyes glued on one spot for nearly six hours! I hurried to open my shop for the brief remainder of the day.

One sunny afternoon a young man came into the shop and we fell to talking about psychic experiences, UFOs and related subjects. He told me he and his friends had been goofing around in the mountains west of town one night at about 2:00 A.M. and had come upon a UFO landed in an open area. They watched for a moment or two, then scrambled out of there, frantic and scared. They were trying to get up enough nerve to go back and see if it would come again.

Although I was quite interested in UFO phenomena, I reacted like most people would to this young man, a fellow right at the "goofing-off" age, and his story—I didn't believe a word of it!

Then we got to talking about psychic experiences. For some reason unknown to myself, I showed him the pendulum and how it worked. He seemed excited and wanted me to ask questions for him to see if we could get answers. I did not tell him it had become for me much more than a toy, but said I would try to "get something for him." Unfortunately he had an appointment and could not stay just then, but we agreed that he would come back to the shop at 7:30 that evening and I would meet him there. He left a series of questions he wanted answered.

That evening, before he came, I asked one of his questions and got a little narrative in reply. I put this aside and did not tell him about this preview. I wanted to ask the same questions again when he was present and see if I got the same answer. It was a matter which I personally knew nothing about.

He came promptly at 7:30, eager and curious, and we settled down in comfortable chairs on either side of my old oak desk to question our little bit of paper and piece of string.

I had the feeling he really believed strongly in this sort of thing. He was eager and cautious, not too skeptical or gullible, with a well-balanced mind that meant anything could happen.

We asked the first usual and dumb question.

"Is someone there?"

"Yes," swung the pendulum.

"Who is it?" I asked.

"Betty Worth."

The young man grabbed the edge of the desk, very excited.

"Ask her—ask her—oh, ask her where I knew her, what city?"

"Seattle."

He became obviously agitated. "Ask her why she never came to the reception."

I had no idea what he was talking about, but I dutifully repeated the question.

"I was jealous," she said, then added all on her own and a little wistfully, *"You never knew it, but I loved you."*

The young man fell back in his chair and waved his hand, "That's enough," he said. "No more of that now."

I waited a few minutes while he pulled himself together.

"I did know a Betty Worth," he said at last, very softly. "We went to school together in Seattle. She died."

He sat for a long time, thinking. He did not want to pursue the conversation. I did not watch him, for I felt I was intruding on something emotional and private. I played with some papers on my desk and wondered if he might just be up to a good job of play-acting. But why would he do that?

Presently he sighed, somewhat raggedly, and said, "Let's get on with this other thing."

The other thing was a question about an event he claimed to have witnessed in a large city in California. It seemed there was a statue of a famous Civil War general displayed in a park where strange events were taking place. The young fellows would gather there late at night and shout taunts at the general. After a while, in the deepest part of the night, a gray form would slowly emerge from the statue and act as though it was attacking its tormentors.

I did not believe a word of this story, but I put forth the necessary questions. First we asked who our informant might be.

"Your uncle George," said the pendulum.

"I did have an uncle George," said the young man. (But then, who doesn't?)

"Ask the name of the general," he suggested. I don't know why we assumed it was necessary that I do the asking and that his voice could not be heard. We just assumed it had to be that way.

So I asked the general's name and the pendulum swung, *"Jackson Arnold Paige."*

The young fellow gulped again. "The name is right," he said. "Now ask why he gets so furious and tries to chase people."

"Wouldn't anyone if they were so taunted?" I wanted to ask, but instead put the question to our unseen conversationist. Back came the same rambling story I had obtained in the afternoon by telepathy. (I did not tell the young man about the telepathy part.)

It seemed the general's mother had "Negro blood" and since he came from the Confederacy, this was necessarily his most carefully guarded secret. The taunts and jeers flung at the statue, that represented him as a hero, were just too much for our poor general to bear. They were regrettably of a rather obscene nature, referring to his ancestry and the marital status of his parents. The youths knew nothing of the general's life. It was simply some foolish "goofing off," something to do on a hot summer night for kicks. It seems that the kids were hitting the general a little too close to his tender spot and he was responding from beyond the grave in the only way a gentleman and a soldier could—by charging into their midst to do battle!

It took several hours to get all this material out of the pendulum and onto the young man's note paper. We were both exhausted with the effort. He seemed genuinely shaken by the whole experience, as though he accepted it as truth. (Or was he merely "goofing off," having a bit of fun with a rather strange lady who kept an antique shop and talked to a piece of string?)

It was quite late by then, and he left, vowing to check out the material on the general and come back. If he ever did come back, it was when I was not there.

The whole episode left me truly perplexed. Taken at its face value, it might have proved something. This would have meant, however, accepting everything the young man claimed to have happened as the gospel truth. It was all a little too hard for me to swallow. The flying saucer story was the factor that perhaps discredited his veracity the most. It was surely a discordant note in what might otherwise have been a truthful, accurate case.

For a brief time, there was a lull in my fears. I seemed to be making some rapid and important progress in my reception and assimilation of the pendulum presentations.

Then, all at once, everything flew apart. It was as though some awful new force had taken over, or an old one had suddenly gained new power. It was Hweig. He had been in the background, hardly communicating. I had the feeling he had been waiting until I had reached a certain point of awareness, a certain strength of understanding. Or maybehe was waiting for the telepathic communication to come into my awareness.

Anyway, back he came, strong, purposeful, breaking up perfectly good sessions with obscene observations and demands. Again, the terror of demon possession took hold of me.

Somehow, Hweig had gained some new power and new strength. He overrode other's conversations, breaking in with his own and bringing with him a whole new chorus of voices, never named, never really identified. Perhaps they were all secondary voices of his own, invented to taunt and terrify. Somehow he seemed jealous, as though he thought the others were taking privileges with my mind that were his alone to take.

There was never any question as to who was speaking when some of his messages came through. *"You are the most stubborn bitch in your whole world,"* the pendulum would spell out. *"Why don't you do what I ask?"*

"Ask proper things in a proper manner and I will be glad to cooperate, just as I do for the others," I said. "Your companions are much nicer than you."

"To hell with being nice," he said.

Then I realized all this was going on in my head! I was not even using the pendulum. Terror struck. I was positive my mind was being used by some horrible creature, perhaps even a type of demon.

Then came more taunts, jeers and denouncements. Now terror gripped me completely. I was sure I had gone completely mad. What horrible, embarrassing, immoral and criminal things might I do? I fell back then on the religious beliefs I had laid aside previously as being too ritualistic. I found a rosary and, for some reason, began bouncing it up and down in my hands. Something in the clatter of the beads gave me encouragement that the demon was being exorcised. All the time I was saying Hail Mary's like a house afire.

During the "exorcism," two women came into the store. Somehow I managed to calm myself enough to wait on them. It took a truly superhuman effort and I must have seemed pale, disheveled and distraugh. The women kept looking at me strangely as though they wanted to speak of it. I had the feeling they wanted to ask me if I was all right, but they did not and eventually left.

Then the telepathic words were back in my mind again.

"You are mine now. There is no escape."

Again I went back to my feverish prayers, and finally had an inspiration to lock the door, lest someone else come in. Eventually I collapsed on the floor behind some boxes, and only half-conscious, heard a voice like thunder in my mind. It said, *"Leave this woman alone!"*

After awhile a little creeping voice said, *"Ida, who speaks from out of the center of your forehead?"*

"I don't know," I whispered.

"Ida, who in the hell are you?"

"Just Ida, that's all. Just Ida."

Soon, everything began to ease up. The pressures in my head were not so intense, the vibrations of my body were negligible, and best of all, the horrible, insinuating telepathic voice in my mind disappeared.

But the others were still there. When I found myself calmed and seemingly uninvaded, I was just curious enough to pick up the pendulum again. The Good Cheer Man was there, and Frank, and now and then some communicant came in for a short time, but seldom stayed long. It seemed on an experimental basis only.

Before long I found myself beginning to enjoy the comradeship of these more gentle, unseen personalities. I joked, laughed, sang, read to them, asked their advice and pointed out fallacies in their explanations. I did all of this quite cheerfully and without my former rancor.

During all this, there would briefly be some intelligent conversation, during which they would actually make sense. They were, in a small way, enlightening to me.

"We cannot tell you more than you are ready to receive," they said. *"You must be prepared and developed to understand what we want to reveal."*

"How can I help?" I asked, forgetting all about my former terrors.

"Do as we tell you."

So I did. I did funny little exercises for my eyes. *"To strengthen the vision. So we may see more clearly through your eyes."* The thought struck me that there might be daily events in my goings on that I did not care to have some invisible neighbors see—like bathroom things and intimate moments with Bill.

"Don't you think we have any delicacy?" they asked. *"Do you think we are physical brutes like those from your own world?"* This was all pendulum now. No more voices. Several times they referred, and somewhat slightingly, to "your world."

"Are you a world apart?" I asked. "A separate world?"

"Not altogether separate," they said, *"coexistent. Oh, you won't understand. We can't explain now. Be content to wait."*

Obviously, they did not think much of my comprehending powers.

I was lighthearted now, happy and excited most of the time. I sang as I did my housework and bubbled good humor at my customers. I felt I was "getting somewhere" with what I still called "my experiments." Little did I realize then that the experiments were not in the least mine, neither to plan nor control. I was being used by these denizens of that strange coexistent world for their experiments, and the control was all theirs.

I believed I could stop the whole thing at any time simply by dropping the pendulum and refusing to go back to it. But I had the arrogant notion I was discovering a whole new world of psychic meaning and event, and I felt I just had to work it out and see where it led. No one can imagine the depth and height of alternate terror and ecstasy that the forward progression of events created.

One warm, summer evening at home, as I sat reading with my door open, someone spoke softly. I looked up to find a slender, middle-aged man, rather dapper in appearance, standing on my little board porch.

"I was to meet my son here," he said, "His name is Raymond Montgomery."

I shook my head. "Sorry. You have the wrong address. I don't know any Raymond Montgomery."

He took a slip of paper from his inner coat and read from it, then looked at me inquiringly. "It is this address."

"But no one lives here by that name," I said.

He seemed oddly insistent. "My son just had a very excellent promotion. He is a missionary. He is having a group of friends in tonight to celebrate and he wanted me to join them." He looked at me expectantly. What could I say? I had never heard of the name. "Maybe it was supposed to be Eighth Street," I suggested. "This is Eighth Avenue."

He shook his head emphatically. "No. He described the house, the location, the color, even the way the driveway comes in. This is the place."

I didn't know what to suggest. "I'm really sorry I can't help you. Do you know your son's phone number? You could come in and call him."

"Yes, I do have his number, but I won't bother you. I noticed a phone booth down the street. I'll go there and call."

For some reason I was most reluctant to let him go. I had a feeling there was something I wanted to tell him, something very important, only I could not formulate the words.

"You're perfectly welcome to use my phone."

"Thank you, no."

I realized he had not driven in a car and I said something about this.

"No, I'm walking," he said. "I am from Chicago and came in by plane. I'm at the hotel."

Somehow things were beginning to feel more and more strange. Perhaps it was only my extraordinary state of mind, but suddenly I felt panicked that he would leave before I could think of what it was I wanted to tell him.

For some reason he, too, hesitated. Then he pulled out a thin tablet or portfolio from somewhere. I don't know where he had a pocket wide enough to carry it, but produce it he did. He opened it and began to read: "The way of Man has been limited by his lack of knowledge of himself. Let Man study his own soul if he would save the world. All is revealed within to one who searches. The time will come when true freedom, freedom of the spirit, will be known on this earth."

"Good heavens!" I thought. "What is this?"

He closed his tablet and smiled, "Are you a Christian?"

"Sort of, I guess. I'm never really sure."

"Good," he said ambiguously, fingering some sheets of paper that had been concealed within the tablet. For one sinking moment I thought he was going to hand me some awful printed tracts or something, and that would have spoiled everything.

I still hadn't told him whatever it was I felt I was being impelled to say. Panicked because he seemed to be leaving, I blurted out clumsily, "Before the year 2000, humans are going to have to learn how to use a new force coming into their lives, or they will destroy themselves."

He smiled, a little wistfully I thought, and said good night. I suppose he went down the driveway as that was the only way to go, but later I could not recall that I had actually seen him leave.

At this time in my life, I was not in a very religious mood. For my previous marriage I had become a convert to Catholicism. It was something that I had considered for years, but needed an impetus to bring to pass. I was quite sincere and joyful in being baptized, but the subsequent events in my marriage turned the whole thing somewhat sour. My husband, for example, when insisting I do something or other, would angrily say, "Remember we are married in the Church. You can't get away from me now. I am your master. I own you."

This put me into top fury, and speaking very softly, as I do when I'm quite angry, I would answer, "The Lord is my master. I own myself." With the breakup of this marriage, however, my religious enthusiasm was in a kind of suspension.

Shortly after the night the man appeared at my door, I began to have strange experiences at night. My sleep would be interfered with after an hour or two of sleep, and all sorts of strange manifestations would take place.

One night I was awakened and scared senseless by the feeling that someone was drawing a sheath-like material through my body, as though it were another skin being drawn between the outer skin and the flesh. I could feel it coming up over my knees, abdomen and chest, rapidly being drawn upward. I sensed several presences on either side of me working hard to draw it further and further. I got the panicky notion that, once it drew up over my face and head, I would be someone else; I would no longer be me. I struggled and fought to move, but was unable to do so. The sheath kept moving up my body, toward my face. I sent out every mental cry of anguish I could devise, but I could not cry out vocally. I was paralyzed, unable to move a muscle or make a sound. Up to my neck it came. It touched my chin. With a terrible effort, I flung myself sideways and cried, "Oh, God, help. Help me. Help me!"

The presences were gone. There was no movement of any kind near my head. My body relaxed and I could sit up. Slowly I became aware of a sensation of a red light being thrown around the room, just under the ceiling. Around and around it went on the side walls, crossing and crisscrossing over my body, and each time there was a terrible burning sensation.

Then through the night a whisper came, *"This is our laser beam."*

Even in this bewildered and frightened state, I wondered, "I thought it was a laser *ray*, not beam."

They said again, *"Laser beam."*

I realized the words were in my ear, a definite vibration of sound, just as though someone were whispering in my ear. It was altogether different from the former telepathy, which was just a movement of meaning, of words. This was an actual vibration of a physical sound.

"It burns," I moaned. "It burns." And I lay in the night, gasping with fear until I fell asleep, exhausted with terror.

In the morning all seemed normal, until I rose and began to make my coffee. Then, all at once, the same kind of whispers as the night before was there again, in my left ear. "Who are you?" I demanded. "What do you want?"

"I am David."

"David? Are you the Biblical David?"

"There were five David's," was the answer.

I considered this seriously. Perhaps there might indeed have been five historical characters named David who somehow became telescoped by time into one semi-mythical character.

David was quite a friendly sort. He stayed with me for some time, and we gradually came to have very pleasant times together. For one thing, he coaxed me to sing. I could not carry a tune and I made it quite clear by demonstration. *"I will help you,"* he said.

There came into my mind a song that had been popular on the radio years before—one I always had a particular regard for:

If I have wounded any soul today
If I have caused one foot to go astray
If I have walked in my own willful way,
Dear Lord, forgive.

Forgive the sins I have confessed to thee
Forgive my secret sins I do not see
Oh guide me, help me, and my keeper be
Dear Lord, forgive.

As I sang, my voice became clearer and stronger, until, after several days of practice, I was quite pleased with one especially emotional rendition. When I had finished, I had tears in my eyes.

"Thank you," I whispered to David.

"You can now do something for me," he said.

"What is it?"

"The psalm you are always reading. Read it aloud once for me."

So I read the 23rd psalm, which I have always thought was one of the most beautiful things ever written. As I finished my very best effort at reading, I added, "That is so beautiful it always sounds like a prayer. I always want to say 'Amen' at the end of it."

"Thank you," said David. *"Thank you, dear Ida."*

During that time, I had a very vivid dream that seemed real. Usually, I cannot remember my dreams in the morning, but this was as vivid as a movie and I was one of the actors. I thought I was a teenaged girl, very close to the time of delivering a baby. I ran, frightened, through a high-walled garden, and there were soldiers, guards perhaps, running after me with drawn swords. I knew that if they caught me, they meant to rip the child from my abdomen. I ran heavily and frantically to reach the gates. As I fell and knew that in a second they would be on me, I woke up. I was panting and sweating and still frightened. This was the second most vivid dream I ever had in my life. It was real. I was there, heavy with child and terrified for her.

It was shortly after this that David gave me the name of Mirhab for a birthday present. This Elysian interlude was rudely interrupted by a second voice, announcing itself in my right ear. It gave itself no name, but it was quite the opposite character of David. It was a troublemaker, a jeerer, a taunter, and I nearly went mad trying to shake it out of my ear.

Some of its reports were just too silly to pay any attention to. They seemed more intended to distract my attention and keep me off balance than anything else. One day it announced that there was a spaceship waiting outside of town to take me away.

Once, years before, while driving between Indio and Blythe, California, in the middle of the night, we saw a circular object rising up from the desert, becoming silvery-white and drawing away very rapidly. We might have taken it to be a rising full moon, but later that night we saw a quarter moon, and there could not be two moons, a full and a quarter, in the same sky on the same night. It was years later before I

wondered if the first object might have been a UFO. I had mentioned
this incident to my correspondents. Now they were telling me it had
been a UFO we had seen and now its occupants had come to take me
away with them. That I could not buy.

"Malarkey," I said to the right-eared voice.

He went on with quite a scenario about what had happened on the
sighting so many years before. According to him, some of the occupants of our car were introduced into the ship that night and all sorts
of weird incidents took place. I refused to listen to him, for I had discovered that by bending my head to the left or right, I could hear the
left-sided or right-sided voices, respectively. So I went around bending
my head to one side, until I got a crick in my neck—but it worked.

At one point, the pendulum entities were pretty impertinent. They
seemed to think that I was in some way connected with a house of ill
repute. I don't know if I was supposed to be the owner, the manager,
one of the employees, or all three. But the questions aroused my indig-
nation to the point that I refused to answer any questions whatsoever.
Having recently read something about the Akashic records (in Atlantis), and not having any idea where the energies behind the pendulum
originated, I somehow believed they must have access to such records.
Each time they presented one of these questions, I replied haughtily,
"Search the records!" thinking that they would find out quickly enough
there that I was not and had never been so connected.

It was not until after the voices started that this was clarified. The
voices questioned what I had meant by the records, and I said sur-
prised, "Why the Akashic records, of course!"

"Is that what you meant?" they whispered. "Each time you said
that, someone went rushing down to the recorder's office."

No one ever explained what their recorder's office was. I thought
maybe they had some way of seeing into the county recorder's office, or
maybe into the police records. It didn't make any sense, but then, what
did?

The voices kept telling me Bill had been arrested and was in the
jail in the city where he lived, a hundred miles away. They said he was
asking for me and that I should phone and ask for his release. It seems
he was supposed to have disposed of me in some manner and if I would
just phone and tell them I was all right, they would release him. Besides my incredulity still being very active, the fact was he phoned me
every day and sometimes several times a day, so I knew he was all right
and in no trouble.

One day, in the little gray house in the alley, I felt impelled to go
through a series of strange gestures. First, I laid aside everything I
wore, even my glasses. (I later analyzed this as a gesture to indicate lay-
ing aside all worldly possessions.) Then I knelt and made a strange gesture with both arms curved over my head, somewhat akin to the horns
of the Egyptian goddess, Hathor. Then I dropped them to my thighs
and rested a moment. Again, I raised them both heavenward, as
though supplicating, and in incredibly graceful (for me) movements,

lowered my arms very slowly, sinuously, to my sides, where again they completely relaxed for a moment. Then my arms were bent at the elbows and I raised my hands, cupped together as though receiving bounty. As my hands seemingly filled, they fell lower and lower with the weight of was being received. A moment's pause. I then lifted them again, still cupped, and turning them over, began to scatter the bountiful gifts, moving my hands back and forth as though moving them across the world. I knew I was making a pledge to share whatever I might receive with the whole world.

Afterward, I tried again to recapture that marvelous feeling of rapture I had during the episode by going through all the motions again and again. But the magic was gone. The rapture was gone. It was just a series of very awkward gestures with no meaning, none of the vital significance that had accompanied the first scene. That first time had been filled with wonder, joy and solemn rapture. The repeated scenes were flat, empty, devoid of meaning. Nothing. I felt, therefore, that the initial episode had contained some vital reality. I knew when I had sworn to scatter the gifts to the world that I was actually and sincerely swearing a solemn vow to do so. I meant it. Then I was told that I was on a kind of galactic television, being watched by many people, from a star in another galaxy called Tea Elsta. I thought this name might be some confusion with telstar from the similarities in the name, for this had been launched rather recently and was still a hot news item. But the answer came back again and again, spelled out, "T-E-A E-L-S-T-A." I had never heard that name before, nor have I heard it since.

I remember walking along the sidewalk to my daughter's house on my 54th birthday. She was having a family birthday dinner for me. I saw a very bright star, which to my disturbed and unnatural gaze, seemed to hang very close above the treetops. It was crisscrossed by brilliant rays, and it was much closer and larger than any star had a right to be.

The right-sided voice managed to whisper that it was a spaceship. It would be waiting on the edge of town for me later. At least I had enough sanity left to know this was a complete falsehood. I bent my head to the left to avoid any further nonsense of this kind. That was when David, recalling the Biblical character of Mirhab, said I might have this name for a birthday present. I accepted it joyfully.

My family knew nothing of what had been going on with the pendulum, the voices, or the strange happenings. I was living in a secret world in my own mind, half-terrorized with fear, half-ecstatic with joy.

I was truly "far-out," and I am sure my family recognized something amiss in my behavior. It must have been quite apparent that I was in some strange state of altered consciousness. I could tell my family was worried when they took me home from the birthday dinner, for they stood about hesitantly with puzzled looks before they finally departed for home.

Then came the culminating phase of the pendulum activity. I still used the pendulum to verify the voices. For a few short days, the pendulum acted as always, making its own ridiculous and often scary suggestions. Then one day it urged, *"Get a pencil."*

This was my first attempt at automatic writing. It did not work. I suppose I was too foggy or impatient. All I ever got was wavy lines.

I had recently bought a different typewriter, and at the urging of the pendulum manipulator (now someone calling himself Rupert), I tried automatic typing. My head was so confused by then that whatever came out was equal confusion, until I cried out in disgust, "This is all garbage. There has been nothing but garbage around here for the past six weeks. I have waited and waited for someone to say something sensible. All I get is garbage."

"It has been all garbage," spelled the typewriter, *"but now we will get serious. There is serious work for you to do."*

But the intervening forces interrupted again. The voices whispered in my ear that I was being watched from the house across the street. When this became too realistic and frightening, I went to stay at my daughter's house for several nights.

In my reading, there had been mention of lost secret records, buried perhaps near the pyramids of Egypt. Actually, they were supposed to be the records of Atlantis, hidden somewhere and the secret forgotten. Once, when the presences told me to ask for something, I had blurted out, "Where are the lost records of Atlantis?" (I had just been reading about them and the question was fresh in my mind.) They did not answer then, but much later, one of the nights when I had become quite frightened at the turn of events and was spending the night at my daughter's house, the whispering voice said, *"You asked for the lost records. We are going to test your ability to think on several levels of consciousness at once. If you can do it, you will know where the lost records are."* And the rest of the night my bed seemed to rock endlessly, or perhaps my head rocked endlessly. I laid in bed and recited, over and over, "Off the island of Crete, in the Temple of Poseidon, at the bottom of the sea." I was not saying this as though in unison with several voices, but in a kind of roundelay. It was a terrible, taxing struggle. I failed and failed and tried again. At last, near morning and exhausted, I thought myself finally successful. But I could not recall the phrases again for nearly ten years!

The next night, the frightening voices took over again. They said there were microphones planted beneath the wallpaper in the room where I slept. The heater, the kitchen stove and the TV were emitting rays that could harm us all, especially the children in the house. Then came ridiculous accusations against one of the family members, who supposedly was engaged in some criminal business and making a secret fortune for himself while his wife and children went lacking.

In the morning, I tried to warn my daughter that there were dangerous elements in the house that could harm them. Remembering the supposed microphones and not wanting to be overheard, I drew her to the kitchen, pointed to the stove, put my fingers to my lips for silence and secrecy, jabbed frantically at the red "on" button and made outlandish gestures. She got the idea that I meant the electrical equipment was attacking us in some way. She tried to tell me there was nothing harmful in the kitchen and the other equipment, but I became angry because she would not listen to my warning. I cannot remember what I said, but a little later I missed her and went about the house looking for her. She was lying rigid on her bed staring at the ceiling.

"What is going on here?" I demanded.

"Nothing," she said, but her voice was very strange.

"Yes, there is. Something very mysterious is going on. Why don't you come downstairs?"

"I will. In a little while."

I now know she was suffering a hell of doubt and confusion as to what was wrong with me. What could she do?

Disturbed again that she would not tell me what was going on, I announced angrily that I was going home. She protested, but I was too irrational to listen.

At home, the voices, as though they had won a battle, celebrated by telling me all the dire things that were going to happen to me. Eventually, they succeeded in convincing me that someone they just called "they" had stuffed me full of dynamite as I slept and I was due to explode any second.

I left the house without a coat (it was November) and headed for the police station to get help. But I could not remember where it was. I went up and down the streets, across muddy lots, into the heart of town and finally into a phone booth. If I could not find the police I would let them find me. Fortunately, I had taken my purse, so I had some change. I told the operator to connect me with the police. It was an emergency. I told the police I was about to be blown up by enemies and gave them the names of the streets where they would find me.

In a very few moments three police cars sizzled up to the curb. After a brief conference, two of the cars left and the officers in the third took me to the police station. I told them I wanted to talk to a woman, because the topic of what parts of my body the dynamite was supposed to be planted in was a rather delicate one to discuss with a man. I talked briefly to a woman in a white uniform, not an officer, while a male officer made a phone call. Then he said, "Come on," and several women stood up from behind their desks as I went through the outer office.

Then there was a hospital, a doctor's examining room, a shot in the arm and a card giving a doctor's name and an hour. I was to present myself to the county health clinic the following morning. An officer took me home, the voices all the time making suggestions what was to happen to me. They told me to show the officer the two microphones I had implanted in my chest, but when I mentioned this he looked extremely nervous and said he had a vital errand down the street and would stop in on his way back. Poor man, he left awfully fast!

I kept my appointment at the health clinic the next morning. The receptionist could not find any record of an appointment having been made. She kept saying, "Who made the appointment? and I kept saying, "It was through the police department." I could not remember the doctor's name at the hospital. The doctor whose name was on my card was in a staff meeting. Would I wait?

I waited for more than an hour and a half; it seemed like days. I went outside and walked around the grounds, checking every fifteen minutes to see if the doctor was free. All the while, the voices told me to go home—no one intended to see me. Eventually I wandered down the street, found a phone booth, called a cab and went back to my daughter's house.

I was tired and hungry and dazed. I had been eating little for weeks, and had forgotten to eat at all for several days. Once, when I had complained to my correspondent that I was not eating or sleeping properly, that they were keeping me much too busy, the pendulum had become abruptly motionless, almost as if shocked. Then it indicated, *"We forgot you still have to eat and sleep."*

Now what in the world did that mean? Obviously "they" did not eat or sleep. What kind of "they's" could that be?

Anyway, I told my daughter I would go back to the clinic when the doctor was definitely in. I do not remember what I did the rest of the afternoon, though I sat by the dining table most of the time. She must have been making phone calls from the upstairs phone.

At dinner, I realized the children had been sent from the house. The atmosphere was still and tense. No one talked. Afterward my son-in-law said, "Why don't we take a little ride in the country?"

I demurred because the voices were telling me an admiral and his wife were coming to take me to a place in the country for a good rest. Then the small portion of sanity that was left to me knew that this was only one more of their never-come-to-pass statements, and I agreed to the ride in the country.

At the rate the car traveled down the highway, I knew this was no leisurely drive in the country. The old van rattled, swerved and roared like fury. All the while the voices told me we were going to be on a TV program in a large city many miles away. For some reason I was a hero and there was a truck right behind us filled with mobile cameras and TV equipment. I considered this quite seriously, but could not think of any reason why I should be a hero and decided it was all as much nonsense as everything that had gone before.

Then we drove through concrete pillars and passed a sign that said "State Hospital." I realized where we were, and I was glad. I wanted to see a doctor at once, to tell him my problems and have him help me get those damnable voices out of my ears.

We sat in a long narrow room that was much too brilliantly lit and waited. A small, nervous man presently darted into the room and behind the counter. My son-in-law rose and spoke to him briefly. The man took some papers from a drawer and approached, smiling insincerely and, I thought, a bit fearfully, as though I might spring up suddenly and shout "BOO!"

He started to ask my son-in-law my name and address, but I interrupted and spoke for myself. I knew perfectly well who I was, my address, phone, occupation, and all the rest. I was perfectly cogent, and the calmest person in the room.

"Who wants to sign this?" asked the little man, beaming as though he were offering a treat.

"I will," I said. I was quite angry. I knew what I was signing, and I knew that if I signed I could get myself out much more easily when I wanted to, than if someone else signed me in.

The paper said, "...for no less than 30 days and not more than 90 days without a review."

My family left, promising to return the next day with more clothing and necessities for my use. I was guided down a darkened hall into a small office seemingly full of fluttery women in white dresses and voices of steel. At once I disliked every one of them intensely.

My valuables were put into a package and a slip of paper presented for me to sign. The list included "wrist watch with sets." Quickly I crossed out "sets" and wrote in "diamonds." No insurance company was going to replace my diamond watch for one with "sets."

Either the voices were quiet then or I did not notice them.

One of the women said she would show me to the bathroom and my bed. I asked to see the doctor. He would not be in until morning.

"I want to see him now," I said.

The woman said she was going to give me something to help me sleep. I said I would take no medication without the doctor's orders. She said they had a standing order to give new patients this particular drug, and if I refused to take it they had ways of making me. I could see no reason in continuing to resist. I had made my point. If the medication was wrong, now the complete responsibility was hers.

I went to bed and to sleep.

In the morning I awoke to find three persons standing beside my bed. One woman introduced herself as "Supervisor" somebody and the man as "Doctor" somebody. The other woman was an attendant, though they called her a medical aide.

The doctor said, "You wanted to see the doctor?"

I looked at him and went back to sleep.

In my five weeks' stay, this was my only interview with a doctor until the day before I left. There was something called "group therapy," in which the same two persons always got up and said the same two things over and over. Everyone else sat and shuffled their feet and held their lips pressed tightly together.

But that first morning, when I awakened and dressed, I sat at a long table and tried to orient myself with the place and the situation. No one paid the slightest bit of attention to me.

Then the voices began again. *"Do you see that door down at the end of the hall?"*

I did.

"The administration office is on the other side. Run through there quickly and tell them you want to go home."

"No." I knew this would be a very poor maneuver and would only get me into trouble.

For several days the voices were there, cajoling me to run through the doors. Then, one day, there came the powerful voice again that sounded strong and authoritative in my mind. It said again, loud and clear, *"Leave this woman alone!"*

Silence. Then the faintest whisper, *"Ida, who are you?"*

Back in the days when things were merely silly and fun, both the pendulum and later the whispers had asked many times, "Who are you?" I could never understand why they should ask me this. Why should it be necessary? They had chosen to communicate with me. They must have some practical knowledge of me. They said they heard through my ears and saw through my eyes. Or had that only been Hweig? No, they seemed to follow my actions every day, every minute. How could I be hidden in any way to them? Surely I was not as hidden to them as they were to me? I had tried to answer this question, "Who are you?" with my name, address, facts of my life and family background, but they had protested, *"We know all that. But who are you?"*

They asked so many times and were never satisfied with any answer I could think of to give them. Finally I made up an entirely facetious phrase which I simply repeated over and over. When they asked, *"Who are you?"* I would say, "A humble person of good intellect and great virtue." It was a truly egotistical statement, said with tongue in cheek, but I did not know how else to answer their question.

And now again, *"Who are you?"*

This was no time for a facetious answer. "Just Ida," I said. "That's all I know. Just Ida."

After this, it was probably a day and a half before I realized the voices were gone.

When Bill came to see me, appalled and distraught, he asked what the doctors called my condition. I did not have a chance to ask them until I was ready to leave, five weeks later.

"Paranoid Schizophrenia," they said.

So that's what this is called!

There was no more pendulum work or telepathic writing for a long time thereafter. They told me the kind of writing I was doing was not automatic, but telepathic, which is quite different. Now I was afraid to pick up the pendulum again for fear it would allow the voices to come back. I could never go through that again!

Six months after leaving the hospital, Bill and I were married.

It was several years before I got up enough nerve to pick up the pendulum again, and I quickly dropped it. I was too frightened of what might happen. Several times over the next four years I tried again, only to drop it quickly. About six years after the first episode, I picked up the pendulum, only to have it spell out, quickly and emphatically, *"Ida...quit!"*

It was another three years before I came back to the pendulum and found enough courage and emotional equilibrium to make a sustained effort "to make sense out of it all."

By this time, Bill and I had moved back to Portland, Oregon, where we each had an antique shop. It was November 1977, nine years after the first episode ended. Little did I know that I was about to embark on what would prove to be the biggest adventure of my life....

PART 2

The UFO Connection

Compassion as a Saving Grace

One day, I absently picked up my old silver cross again and began using it as a pendulum. It began to swing wildly, as if beckoning me to continue. The message was blunt and the point. *"Get a pencil."* Over the next several evenings, it kept giving me this same message. Finally, just for fun, I *did* get a pencil and poised it over a piece of paper. Not much happened.

Several evenings later, the pencil slid over the paper in dips and waves and I just let it move freely. For a long time, nothing appeared except circles, waves and humps. Over time, however, letters began to appear. They were all running together in one great long word, but it was clearly a message.

I noticed I seemed to hear the words in my mind before I wrote them down. When I asked questions, the words told me it was *telepathic* writing and not automatic writing at all.[1] I kept on with this until I did not even need the writing unless it was something I wanted to keep for notes. I could call on my communicators any time and know what they were saying, in my mind, and could send back messages without speaking.

It was on May 12th, 1978, that the telepathic writing first began to make sense. [Editor's Note: For easier readability and distinction, all communication from Ida's other-worldly friends has been indented and set in a different type style.] First, there were indecipherable lines, then words all run together without spaces, punctuation, or capitals. It was surprisingly hard to read this way. Punctuated, it read:

> …clothing is not the best way to be presented to many ideas in absence of thought.…

That almost says something, but I cannot decide just what. Then, these thought came to me:

---

1. With telepathic writing, you hear the words in your head and write them down verbatim, just like dictation. Automatic writing is when your hand just moves across the page, and words just appear on the page. You don't hear any voices, and you're not aware of what you're writing.—Ida

> Be prepared to write many things you do not understand now. You will in time come to see what we are trying to do. You are improving, but do not question our purposes. We will not harm you nor take over your personality. You are *you*. We only use your mind and muscles. We do not compel obedience. You are guarded by powerful friends who watch over you. Be assured your ambitions will be considered.
>
> We are pleased with your progress.

Here, I evidently questioned whether I was doing the writing entirely on my own. *"No, it is we...Jamie. Jamie. Jamie."* The voice continued:

> You are deciding for yourself that you will help us. ...Yes, it is for good causes. Answers will come soon. There is no way to hurry events. Be patient. We are attempting all possible ways to get you free.

Now what did they mean about freeing me? Bill and I were wanting to sell both of our antique shops to have more leisure time. Perhaps this was what they were alluding to. This had been planned before the return of the "pendulum boys."

> You will soon be working on something more interesting. It may be hard to understand at first. Keep all notes. Later you will know how to use them. Be certain we are going to use your special gifts to the very best purposes. You have been prepared for long years for this task. We will contact you whenever you can find the time and opportunity without someone watching over you so closely.

Again, I had told no one about my experience with the pendulum and this new development with telepathic writing.

By this time, mental telepathy had strongly developed in my mind. "They" could talk to me at any time by merely thinking I was opening my mind for them to receive my thoughts. I did not like this invasion of privacy. There was no way I could shut my mind off to them. They knew everything I was thinking, and conversation went on constantly. The written messages only backed up and clarified the dialogue that went on within my head. On May 14th, I made a note that they had asked me not to question them about spiritual things. During the first pendulum episode so many years before, my communicants had very early on told me to not ask about divine matters. Now I asked, "Who are you?" A powerful message vibrated powerfully through my mind:

No man or woman may know us!

I tried to debate with myself whether or not I would go on writing for them. Who were they; what did they want me to write? In a hastily written note to myself, I said, "I have been trying to do all the things I think Mother thinks I should do, and all the things I think Bill thinks *I* should do. I call this love of others. I am not doing all the things I think I should do, and I call this self-sacrifice. But I don't think I have the right to do that. I think I have the responsibility to do 'my thing' also.

The telepathic contacts continued, and I religiously kept journals of my experiences. These journals are as follows, verbatim, with only minor modifications applied to ensure clarity.

May 15, 1978

The feel of a powerful something, a presence perhaps, interrupts my reading. My head fuzzes and there seem to be sudden jerks or thuds to the bed on which I lie, but not really physical jerks. The powerful personality that was present yesterday when my communicants said, *"He is here,"* seems to be present again. *"He is taking hold of your life,"* they said then. I lean against my backrest and close my eyes. The fuzzy feeling in my head continues, particularly between my eyes. In this area, there is a sensation as though someone is moving all the furniture around in my head. There is a strong point of concentration, a sort of inward puckering, in a spot near the center of my forehead. I wait. Nothing more happens. I take up my book and continued to read.

May 17, 1978

Today the voices said, *"We are not going to see you for awhile."* Tonight I notice my head is lighter, not so constricted. No one is "moving the furniture around in my forehead." I feel lighter, happier, not so compelled or driven. I have no idea who they really are. I am sure they are not disembodied ghosts. I am sure they are not my subconscious. They seem to have purpose and intent that these entities would not have. Who then? *"No man and no woman may know us,"* he said. What do they want of me?

"We are not infallible," they said, *"Do not be angry."* That was when they promised something would happen, and it did not. I was only angry at them for misleading me, not for telling things straight. Was it all a test to see my reactions, to see how far I would let them go before becoming angry? To see my level of tolerance? Would that serve any propose to anyone? They seem to know my background very well; why do they not know my psychological makeup as well? Or were they only testing and developing my receptivity?

"We are not ready to reveal more," they said. I feel it is not that they are not ready, but that they feel I am not ready at this time. They have been waiting a long time for me to get ready. But why me? They are not portions of myself, or an over-personality existing in dimensions other than the space-time physical world as we know it. They are no part of me and I am not part of them. That I know by the directions they give me when that powerful one comes, such as *"Be wary. Do not be greedy. Do not ask questions."*

Last night he was close by as I wrote. Tonight he is not here. I can tell by the pressure in the atmosphere in the room.

A few days ago, I asked, "Do you know what I am thinking even when I do not hold the pendulum?" *"No,"* they signaled. Today they responded to something I had been previously thinking as soon as I picked up the pendulum. "So you do know what I am thinking other times?" *"Yes,"* they answered.

It is this sort of thing that angers me. Their answers are inconsistent, and therefore, in my eyes, irresponsible.

Yet they do give me excellent advice—do this, don't do that—and, as I study it and compare it to my own thinking, I can see their advice is the best thing for me to follow.

Sometimes I think I can detect just a note of derision, a kind of sneer in their messages from the pendulum. How in the world can one detect tone in the swinging of a paper clip or safety pin? I don't know. All I know is that I do. This, however, has not happened much lately, not since he appeared and there has been mental telepathy as well as the pendulum. Sometimes the messages are cut short, as though interrupted by instructions from another source. The pendulum stops dead in the middle of a sentence, then they say something oblique—not at all what they started to say. I feel that "they" neither know me very well, nor understand what makes me tick. Yet I feel he knows all about me.

"We are trying to be patient with you," they said.

He is patient as eternity, waiting. Waiting for me to develop to a certain stage of awareness, to stop fussing about trifles and demanding attention, to be ready to settle down and listen quietly, and to be able to make intelligent responses, ready to stop soaking up the rich benefits of the universe and to start giving out—what?

Tonight there is no excitement, no tension to communicate, but only a quick falling to sleep whenever I try to read.

A real change came into the communication then. It was as though some preliminary work had been accomplished and new work begun. Or had there been a change in correspondents?

There was some written discussion about a sale I was supposed to make, which would enable us to go out of business in style. This discussion went on for days with the constant assurances the sale would be consummated. It almost was, then it fell flat. I asked them to stop reassuring me it would materialize. If it did, fine. If it did not, I did not want to be annoyed at them for misleading me. In other words, I asked them to stay out of it. We went on, then, to other matters.

> ...No, we are not departed spirits. We are energies also, but not as Seth describes it in the Jane Roberts books.[2] He simplifies, but from another area of being. Electromagnetic energies are used, but not as they are understood by your kind. Later they will know more, some through your writings as dictated by him.

Here I asked if I might explain what was going on to my husband, Bill.

2. Roberts, Jane. *Seth Speaks*. Englewood Cliffs, NJ: Prentice-Hall, 1972.

> Yes, if you think it wise. We understand. It is best to wait. Yes,
> when you feel the time is right you may tell Bill *only.…* Yes, it is O.K.
> to ask, but we cannot promise to answer everything.

One night as I sat in bed reading I had the sensation of everything stopping in my mind. It just froze, and became blank. I could not read or think. This lasted for several minutes, then I said, "Let me out of this. I don't like it." And when my head felt useful again, I said, "What were you trying to do, hypnotize me?"

> Right. And you were not. You are not easy to hypnotize and that
> is good for our purposes. We cannot go on with our work until we
> know you cannot be controlled by a hypnotist. It is doubtful if you can
> be. Later we can conduct experiments of many kinds, but not until we
> have more undisturbed time.

Here, for the first time, I brought up the subject of UFOs.

> They exist and are not to be taken lightly. They pose no threat at
> present. We can only say now they are being surveyed by Earth sci-
> entists, but secretly, and they soon will be better known. There exist
> excellent photos in several European countries and in the United
> States also. In the coming year much will be revealed. Some are man-
> ufactured objects of physical elements, some are not. They originate
> far from your planet, on several planets outside your solar system.
> They are based under the sea here on Earth, but originate elsewhere.

I asked about abductees.

> Yes, many have been taken, some returned and some carried
> away. Many animals have been taken away. We are speaking true
> now. Before we had to test you in many ways. Why don't you take
> time to read over what we have transmitted so far from the beginning.
> We want you to write now about your first experiences with the
> pendulum. Try to recall as much as you can. We would like it to be
> done by the end of the year. Leave off just where writing was intro-
> duced. Good-bye now…*Amorto.*

My correspondents seemed to take much interest in the trip to Hawaii Bill and I were planning. Demands were made:

> Answer this question: Where are you going in Hawaii? Go to the
> park where the porpoises are. That is excellent. …Yes, we need to
> see them through your eyes.
> One of us would like you to see what is called the "Heavenly
> Grace." …No, it is not a church; it is a palace.

I asked if he meant the Heavenly Bird, which is the Iolani Palace.

> …Yes, it is the Iolani Palace. We cannot all be with you all the time,
> but we can come from time to time to see what is going on and to see

things we would like to. ....No, it does not work that way. We need your physical eyesight and brain to let us see what we wish to.

...No, you will not be noted for a nut before you die, or after either. You will be noted for a good writer and psychic.

Of course it will be published; that is the whole reason for writing it. Sometimes you are very stubborn. (I think he meant obtuse.) We will try not to frighten you again. What outline? (Here I must have mentioned an outline for a book I had written some years before and had buried in a trunk.) ...No, we do not know about any outline. What is it? ...Yes, we can [check up on it] and we will talk later tonight. G'bye...*Amorto.*

Another voice chimed:

Only a few can understand the meaning of your outline now. In a few years, other works will point the way. The time is not yet—soon. Meanwhile there is much other work to be done. Hawaii will make you more relaxed and rested, then we begin. Plans change from day to day as probabilities are chosen.

Very soon there will be important work for you to begin. For now, work on the exposition of the way in which you learned to use the pendulum and the results, including the first bad try of nearly ten years ago. You can write a very entertaining account of this. Be humorous if you wish, but not facetious. Do not obscure the fact that it is real and important. Not all people can use it, but many can. Many will be helped if they try accepting the truth of it and work against being too skeptical. An open mind is absolutely necessary. Wariness, yes, but not hard skepticism. The forces and principles will be given at a more proper time. For now just the experience is to be told.

May 29, 1978

All the time I was awake, we fired questions and answers back and forth on many subjects—all day long. Observations were made on both sides. Some fearful and some very funny interchanges took place. It seems everything they particularly wanted remembered, they told me to write down. It is all in the notes.

We were working with new tools last time and none of us was adept enough to block interferences. We lost control to other mischievous (not malicious) ones. Everyone was appalled when you became ill and a halt was called to such activity for a long time until you were again emotionally ready and we were better practiced in our part of the work.

You are safe and protected and we know now how to keep you so. Be not fearful. Take no heed of warnings from outside about the dangers of this. You have progressed beyond danger. You now realize how to avoid some of the former pitfalls, and we have learned how to protect you from ulterior influences.

After the first few words of the next paragraph, the single long word separated into the individual words and never again did I have to struggle to read what had been written. Handwriting styles had changed again and again, but this one was much like my own and stayed that way.

These next half dozen paragraphs took place in a shopping mall. I had been trying to decide whether to buy a new dress or just buy material to sew one. The inner voice asked me to find a place to write for a moment.

> Don't get involved in a big job of sewing. You have better things to do now. Buy the dress!

After shopping, I went to a restaurant in the mall for lunch and there was asked to write a note.

> Your new dress is truly charming. But *you* are too chubby, shall we say? Your scale is not exact. It goes this way and that. You did lose a trifle, but keep at it. You have to eat something, but go on with your resolution to cut out candy, etc.
>
> You are getting conservative with your money. We did say don't give your money away, which you are inclined to do! Very properly done! But don't deny yourself the things you need or even don't need and just want, within reason. ...Yes, a new typewriter with fine aesthetic type will make you feel better about the work involved. You will feel more creative if the print is nice.
>
> We are getting you down pat. I guess we know you better than anyone has been known on your "side of the fence." At least by us. It is only because you are so agreeable that we can. We know your limits and we will not trespass. We don't want to lose you. We have invested a lot of enthusiasm and hard work.
>
> You are now being loaned. Later, you will work for your protector, after you have established yourself with our writing (which is not so complex). There is a connection between us and our way of life and some of the revelations in your outline of so many years ago.

At the beginning of the second era with the pendulum, it had told me two deceased members of the family were waiting to talk to me. What they said was not important, but I now asked if they had really been spirits of the dead, or what?

> I can tell you this much. They were not true spirits of the dead. Not their true personages.
> ...Yes, a reconstruction, but too difficult to describe.

Here I informed him that I had been able to tell there was something rather phony about them.

> You make us hang our heads. We thought we were being so subtle. But the messages were truly given.

May 30, 1978—9:45 A.M.

After a little while you will not need a notebook for this at all. It will all be direct telepathy, but this takes much practice so you can distinguish your thoughts from ours. Always be prepared to take notes on what you want to remember. We can use a coexistent and coextensive universe—a different dimension of being you would call it—although that is not altogether the correct term. This occupies the same space, but in a different concept of time. ...Yes, you are right, if it had the same concept of time it could not coexist. If you could manipulate time and energy, you could come into this area of being, as we do. ...Right again, when we use this coexistent world, we are as an energy existence—that is, from your viewpoint and understanding. We are not, then, material, as you are. We are trying to think in new terms, translating for your understanding. We then exist in a different kind of time and know how to use energies in different ways. There are more kinds of energy available to you even more than you have discovered. You are just beginning to catch a glimpse of psychic energies. No more now; we have to translate this kind of knowledge into your understanding. It is like a foreign language. We must study also.

You just did it again!

I seemed to "just know" that they had become more solemn than before when they were laughing and joking. I could not *see* them in any sense of the term, but somehow I thought I could detect what was going on. Of course, they could have been just kidding me when I reported such things.

—1:06 P.M.

Not right now.

—1:17 P.M.

Partly that.

—1:54 P.M.

No, please don't interrupt now.

—4:11 P.M.

We had quite a meeting and found some answers to questions. Contact us before bedtime and we can reveal some things learned and some events to come. No, not sad, but chastened. We were intruding on someone else's plans, but it has been worked out cooperatively now. You are not to be taken so lightly! Your purposes are far more serious than we knew. We know how confusing all this must be, but patience! It will be all the better for waiting. You can do well what we are proposing. It will be writing. Keep working on your story. We will help on details. We were otherwise engaged this afternoon. G'bye for now....*Amorto.*

—11:35 P.M.

We have been in conference all day with others and have discovered we were not alone in working with your abilities. That is why they have developed so fast. But the others were working along different lines, much too complicated to explain when you know so little. ...Yes, you can tell what they are doing if you are told to expect it. ...Yes, we will try to tell you what to watch for.

The personalities who converse with you cannot be named yet, and actually the names would mean nothing to you. You have never heard any of these names.

You do have physical sensations when they direct certain energies at you. The vibrations of limbs is one. We did not know about that. The tightness in head is because of us, for we direct our energies there. You are indeed being energized all over, a good way to put it. They also direct energies to your physical senses. ...No, not like visions. ...Yes, that is why you seem to see without actually seeing our room, why you thought you saw a man with glasses, and why you know without either seeing or hearing when we are laughing, smiling, joking, serious, etc.

It is indeed a relieved kind of joking now, and sedate. We were severely chastised and scolded for infringing on your thoughts. That has become forbidden territory. Now we can communicate with you only when you direct your thoughts at us, and we can answer only by this telepathic writing. You are right, it is not automatic, it is *telepathic* writing. But you have to pick up the pen and be ready before the words come into your head. We did not know ourselves just how this was being transmitted. Just a different faction. We did not have as much right as they, for we were under-using your potentials. They have been training you much longer than we. ...Yes, your guardian has known, but his position is such that we are not in communication with him. There was no real harm in the double training—he was watching carefully. ...Yes, he was rather amused. We were working so hard and rather unnecessarily. The separate works are all being amalgamated so we can combine on a cooperative basis. It will be more relaxed for you. ...Of course you are confused and it all seems completely unreal.

You will have to excuse us for being so condescending. We did not know all about your past training and present status. I am indeed the leader in our group, but even I had no knowledge of this other work. How to describe—I send for research answers and get only as much as *I ask*. I did not ask the right questions, therefore was bereft of all the information I should have had.

Be alert to all that goes on! Alert!

Do we have your permission to read your entire biography?

Here I answered, "You have my permission, but don't remind me of everything you read."

...O.K. We won't remind you of what you would rather forget. Neither are we proud of everything in our lives. We are not angels. ...Yes, and we are extremely pleased with the way your story is progressing much faster than we had hoped.

We want to get something practical across tonight. It will help verify our reality and good intentions.

Make a good beginning by understanding our position. We are teachers only, not miracle workers. When you saw something like a school or university, you were close to fact.

I am indeed a leader, or like a head professor, in this group. It is not like a religious or political or hired group. When you compare it to a fraternal organization, this is the closest, though not precisely that either. We try to help persons like yourself find abilities to cope with some of the more mysterious and unseen forces surrounding you. You are blind to many things that affect your life; we try to help you see. Humans were not always thus blind. They have lost their abilities through misuse and disuse. Those societies that do not forbid such work still retain some ability, but are greatly diluted by religious taboos and ignorance and fear. These had to be overcome in you also, and are not yet laid very securely to rest.

...By all means, use everything and anything we write here to whatever good purpose they may be put. Maybe we have already started "Book 1."

You guessed we were working in retribution for things either left undone or not properly done. Our task is to help beings on other levels or in other areas of being. We are not angels! Perfection is not one of our attributes. We are compassionate and that is our saving grace. (It is your saving grace as well.) Indeed, a person's worth is measured by his or her compassion. You are so right! Our books will be written with the greatest compassion for the problems and troubles of humanity. Now which of us actually said that? We are beginning to think together, but only when you hold the pen.

We are working together well. The name on the book as author *must be yours.* There is a very vital, very important reason for that. The works are what is important. Your name as author will be used. Explanations later.

It is too technical to discover the layers of being and non-being, substance and non-substance in our existence. How to put all this into your terms (when you have no terms for much of it)? We have to use the terms you know and understand to translate our meanings; therefore, it sometimes seems like something you already know. We will get around that presently with more practice. This is difficult—to stretch both our abilities into new channels. You help with the translation more than you realize. That is why so much of what has gone on before seems to be superficial talk. Technicalities or technical explanations slow us down, due to the problems of translation from our jargon to your language. A mystery is revealed there if you can stretch your imagination enough to catch it!

Please relax your mind. You thoughts are mixing up with my thoughts.

June 3, 1978—2:55 P.M.

...Yes, I wanted to talk to you. We are going to experiment a little on your receptivity. Let us know what you detect now.

Somehow I just knew he was nodding his head.

...Yes, I did nod. How did you know?
Now what are we doing and how many are here? Three is right. How did you know? Now what is going on and how do you know? ...No, it is not ridiculous. We do have brows. And he did jump up, pace the floor, wipe his brow. What? No kidding, Ida. This is terribly serious.
...Yes, two factions. We are identities. No more now. We have a lot of thinking and conferring and research to do...*Amorto.*

—7:20 P.M.

The conversations in my head were now continuous. I had no time to really think, put two and two together or even go back and read my notes. If I were not actually working in the store, talking to a customer or talking to someone at home, the voices went on and on in my mind. They were very light and not vibratory—there was just a meaning going through my mind, as though I were talking to myself. Sometimes it got mixed up with my own thinking, and I would demand, "Who said that?" I never quite knew if it were my thought or the others'. It was not at all grim—in fact, it began to be quite fun, as soon as I got over being scared to death at the happening. We developed many jokes between us. Sometimes they asked questions I did not want to answer, then I quickly said, "Censored," and we both laughed. I tried to keep my response to the question from coming into thought, but I never knew just how successful I was in stopping it. I never knew if it was there adequately for them to guess what it would have been or not. I'm afraid that sometimes it was.

When I said, "We both laughed," that was exactly what happened, for somewhere along in here there began to be a kind of spasm in my abdomen that could have been the same as the movement when a man chuckles or laughs. After this had occurred for several weeks, I maliciously pointed out that only men laugh from their abdomen, while women laugh more from their diaphragm.

"Ida, you stinker!" was the response, followed by such a hearty belly laugh that I clutched myself and said, "Hey, that's too hard!" Often the chuckle was followed by a deep sigh, but I did not call this to their attention until much later.

...By all means. This has been a revealing session and we can now organize our work on a stable foundation without running into so many surprises and sudden changes. We do know a lot more about you than you know about yourself!
Go watch TV. Come back after wrestling...*Amorto, Jamie, One.*

Since my unseen friends were, at all times, seeing through my eyes and hearing through my ears, they also were watching the TV programs with me. Some they could not stand. The commercials...well, I always had an impulse to shut my eyes. But the wrestling they loved, and sometimes I had to beg them to relax, because the tension of watching the screen so hard gave me a headache. After TV this night, I went back to my notebook.

> There is too much interruption and the time is late, but Amorto wants to talk.
> We are analyzing your life and finding a coherent pattern through-out relative to this development in psychic abilities, as though you were directed from the beginning, yet always within the choice of free will. ...No, it was not just an accidental development. It was carefully planned and guided from the beginning. Everything fits too neatly into the proper place at the proper time for it not to be planned.
> All this is very exciting to us, as we do not often have the opportu-nity to survey a whole lifetime. Yes, down to the minutest minute! Ac-tually you have had a rather exciting life, and we envy many of your experiences.
> There is no measuring what we have accomplished today. It is ore than all the past work together since we have "met" you. You have saved us countless days of rearranging and struggling to keep things coherent. Thank you for allowing us this possibility. We will not make you sorry in any way that you gave us this opportunity. We think you are a wonderful gal!
> It is *all real* and so are we and so are you! Good night...*Amorto, Jamie and One.*

About this time I remembered and wrote into my pendulum story the account of the strange roundelay concerning the lost records of At-lantis. Just as I learned it, I wrote, "Off the island of Crete, in the tem-ple of Poseidon, at the bottom of the sea." According to my informants during the first pendulum episode, that was where the lost records could be found. As I wrote this into my story, I was astonished to find my current correspondents in wild jubilation. And at their insistence to let them write, I received the following:

> You don't know what you have written! A priceless, priceless trea-sure! The lost records of Atlantis. You remember all right! Exultation! We are indeed laughing and weeping at the same time. You are O.K. It is all real!

It certainly seemed too far out to me to have any reality. I cannot estimate how many times after that I was to ask suddenly, "Was it re-ally true about the lost records?" And each and every time they assured me over and over again that it was indeed true.

Now there were some changes made in my correspondents.

> Someone else is going to take over. You will be well taken care of, have no alarm. *One* will continue as always, but Amorto and Jamie must go. Suma. Suma is going to be your correspondent. And *One*

and somebody else not decided as yet. So sorry if this alarms you. Please don't be alarmed. (I was getting a bit panicky and they well knew it). We are just disappointed that we cannot continue. We are so fond of you. He will speak to you now. Good-bye, Ida. Don't forget us. Yes, from time to time. Please don't be alarmed...*Jamie and Amorto.*

And then a deep voice introduced himself:

You are not to be frightened. I am Suma. You spoke to me once, long ago. You have been unhappy that our correspondence seemed only on a chit-chat level. It will be so no longer. Rest assured you are protected at all times. We have no means of harming you. That [the first period of using the pendulum when they had frightened me so] was a mistake. It will not happen again. We know that you are not happy with our practical progress, but we had to become organized or have a well-structured plan of operation first. That is ready now. We can all proceed with some useful work.

Seeing Through the Veil

June 4, 1978—7:00 A.M.
This morning, I was awakened rather abruptly:

> We are showing you off to, well, not our alumni, but to former as-
> sociates and fellow workers. ...Of course, not *everything,* goose. It
> makes us the men of the hour. Honest! You cannot realize how elat-
> ed, and how proud we are of our part in discovering you, even if others
> were helping in many ways.

Now what did all that mean?

> So, Ida, you are seeing things again. Please describe our room as
> you see it. ...No, that is not ridiculous. It is something like that, but de-
> tails are not good.
> Yes, room described O.K. ...Long table. O.K. ...Men described,
> not right... Things, tablets and pens, O.K. Skip people for now. Num-
> ber is correct; five present. ...Stunned is the word! ...We can't answer
> that now. ...Side room off courtyard, yes.... Yellow cream stucco,
> yes.... You are describing closely, but not exact.... There is such a
> desk; utilitarian is right.... Enough for now.

—8:20 A.M.

> Thank you, we have been eagerly waiting. Our friends here want
> to ask some questions. ...Yes, this is Roger.

I seemed to have heard, without actually hearing, that someone
had called him, *"Roger, you old dog!"*

> No! In what manner did you hear that? All right, thank you. Now...
> why are you so much alone?

I prefer it.

> Very good. Why are you so fat?

I guess I eat too much.

> O.K. Can you cut down?

I'll try.

> Why are you not more active in social groups and others, like political groups?

My mind is centered on other things.

> ...We understand. Good. You have all the right answers. One more, then you may go. Do you truly love your husband?

Why else would I have married him?

> ...Yes, very good. Thank you. Do your thing. We will not intrude on your mind. Be free. When you can this afternoon, give us some time. O.K....right.

—9:37 P.M.

> We have been waiting, not too patiently, to see you. Some of us have prepared a questionnaire. Will you be so good as to answer? Let's begin.
> Why is there no manner of excitement in your life? No dinner parties and going out and so on....
> Are you ever going to be a grandmother again?

"What a strange question," I said.

> Good. ...Yes.
> Why are there.... Excuse us, skip it. We have to confer.... Make your phone call, then come back.
> There is a little misunderstanding on our part. Do you have only one daughter? ...Yes, does she have children?...All right. What we wanted to know was, how many descendants you had altogether. ...Yes, it is important to us. ...Yes, thank you.
> You have some questions and we have some explaining to do. Quite a lot, in fact.... First?

Here I spoke of a TV program we had watched about the coming of a UFO in which the symbols of "The Flame and the Rose" were displayed with a great deal of mystery. I asked them about these symbols.

> We don't know either. But if that were a true event, there are tremendous potentials of evaluation. Some things we may all learn. If it was fiction, we will waste a lot of time. We are going after it....
> True. They [the UFO people] did live in this world once, eons and eons ago. ...No, much longer than Atlantis; they were the first altogether "human" people! Now they live on a planet in a far galaxy. ...Yes, some of the UFOs are hardware, of a metal like iron, only much harder.

"Are they like us?" I asked.

> Somewhat. ...Yes, they would like to return to Earth, but have developed in a different environment. They are testing soil, atmosphere,

etc., to see if they could survive here. We don't know how many there are. Gospel. ...No....*real.*

We cannot prognosticate coming events any more than you can, except by probabilities at which we are adept, but this allows margin for miscalculation. We are not infallible. You have been so uncommonly kind in giving information and not becoming too angry at impertinencies that do arise from time to time—we apologize for the "fat."

There have indeed been a tremendous amount of events since we last spoke. Your information [on the lost records of Atlantis] sent us all into the seventh heaven of delight. It was a secret we had been trying to uncover for uncountable time. I know it is impossible for you to grasp the reality of it, since you know so little about us. Never mind the details of that now. The secret was real enough, and enabled us, through "remote control," to ascertain the truth of the repository of the records. They are there, and will be found by humans in time. Meanwhile, we have our ways of deciphering them. ...Yes, even if we cannot actually see them. ...Very complicated...could not even begin to explain. They are there. One thing we are not able to discover: Who told you that secret and why you? ...Now wait, women are not supposed to be logical.

June 6, 1978—1:26 P.M.

The Betty Book[1] is so close to our reality. So very close! You must read it sometime again, but not right now. The Castaneda book will amuse you, but don't take it seriously. It is mostly farce. *The Betty Book* is the closest you will read in English. The Russians are far in advance over your country. One of the reasons you must be ready for future changes. Secret maneuvers. Inner invasion is right!...*Roger*

June 7, 1978—2:20 P.M. (at the shop)

We have a lot to say! You did not hear us, yet you obeyed our request to go to your desk. Right. ...O.K. We know, so don't try to remember. Do you "hear" us now? ...All right. Can you hear this? ...Now, can you hear this? Put your pen to paper and look away.

This kind of "hearing" they were referring to was not at all the mind message. It was no kind of message at all, just a sort of inspiration to do certain things.

I drew a diagram without actually hearing any message.

Very good! You did not hear anything?

They seemed quite pleased that they were getting themselves clearly across without me being conscious of any voice or message. I did not like it at all. I felt my mind and actions could be influenced by them at any time without my knowledge. I was beginning to get cold feet over the whole proposition.

---

1. White, Stewart Edward. *The Betty Book.* New York, NY: E. P. Dutton & Co., 1937.

I told them about an incident during the first pendulum episode. I had seen a very vivid blue flame outline the little silver cross I was using as a weight. On their silent instructions, I drew a diagram like a cross and flame.

> Is that like the flame?

"No," I said.

> What was different? Draw it.

So I drew it as I had seen it.

> Very good. We know what it meant. Great significance. Thanks for insisting…. Be alert to unheard messages. …Yes, like that. You didn't hear the notice to sing? We are so excited that we can't work at our best. Very great significance.

I began to complain. I did not like the situation as it seemed to me.

> You always have free will. We cannot encroach upon that ever, even if we wanted to. …Yes, Suma now. How did you know?

Suma always began his writing by making small circles for a few seconds before letters appeared. So I told him.

> Fine. These are the things we need to know. How we are coming across in not-heard instructions. Suggestions is a better word, but will use instructions as long as you know we are not authoritative about it.
>
> Delivering your book material will be no easy task because of the translation necessary. Our language is pictorial or patterned telepathy for the most part. Though we do have vocal language, we seldom use it because it is so different from your speech or vocal words. The task seems insurmountable at times. But you are very quick to grasp meaning and help with the translation more than you know.

Many times, as one of them was speaking, I could feel several different words with the same meaning "coming up" and I would not know, until the actual writing, which one it was to be. Often it would be a synonym that I had not detected until they used it.

June 11, 1978—7:20 P.M.
"Did you want to say something?" I asked.

> …Yes, for just a moment. We would like to see the animal picture on TV and the other show relative to St. Germain. We will discuss it then.
>
> About this momentary vision: Draw me a diagram or picture of what you saw. …Good enough. We know what you mean. …Perhaps; let us not discuss it now. We will wait for you.

—9:00 P.M.

> Thank you for letting us see the shark picture. The St. Germain picture was all fabricated nonsense. We know of St. Germain, and his life was quite normal, though longer than most. He was learned and scholarly and knew some subtle tricks, but was in no way supernatural or superhuman.
>
> We have had a busy day here, showing off your world to our scientists. …Yes, it is a matter of vibrations, but very subtle and complicated patterns thereof. Later you will learn more about this. This is the first time that others, i.e., other than we few experimenters, have seen directly into your world through the eyes and ears of a "your-world" person. That is indeed a complicated sentence.
>
> We have enjoyed your flowers and presentations today. Sorry to keep you rather busy when you had so much else to do. This is the first glimpse into your world for some of us.

I had a session in the early afternoon, during which I was asked to please go about the yard, pointing out and naming flowers. It seems they recognized what the flowers should be as long as they were told the name. Or maybe if they saw the flower and were told the name they could look up some reference. I could not imagine what the point could be, especially since each time I tried to add some fact about the flower, the "voice" came back rather dryly, *"We know."* Someone certainly did not like to be told!

Later the same afternoon, I was asked to go to the store and point out various fruits, vegetables and other grocery items, just naming each one. I would simply pick up a cucumber or eggplant, examine it slowly as I named it in a whisper, then lay it down and slowly proceed to the next vegetable. They did not want to look at the meat, and, in fact, asked me to hurry past the meat counter.

Later that evening, as I packed for my trip to Hawaii, a urgent voice cautioned me:

> Now something important! Take precaution against this [my journal] being found while you are gone. We are not ready to be revealed yet. Much must be done first to make you ready to answer questions.

Glimpses of Beauty

June 14, 1978 (On the plane to Hawaii)

Prior to boarding the plane, I had walked about slowly and named things as though pointing out items of interest to someone new to the sights. We then watched a movie concerning the ballet.

> We want to thank you for the most wonderful event! None of us now present has ever seen into your world before! This is the Absolute Truth! We know about all these things, but can see them only through your eyes. Never has anyone else given us vision. Truth! Thank you.

I don't mind saying I was a very puzzled person! Obviously, someone new was now observing, but I was not introduced.

> The ballet was *breathtaking* (your idiom). We are glad you explained the story to Bill, as we were confused as to what and why things went on. We never dreamed anyone could be so cooperative. You don't know us or anything about us except what we tell you, and sometimes you realize it is necessarily a shaded or half-truth. It is vital. We are so thrilled you remembered *The Betty Book.* It is superb…. You must take the rest of the day for yourself. We will not "see" you again until you get up in the morning…*I, One, Suma.*

June 15, 1978 (Honolulu)

It is so nice to be able to view all these things. The ocean was exciting this morning. Even the action of the wind (the motion shown by the palms) had a certain charm; like the ballet. I write all this in a most simple style for a purpose, just to see if I could still write my own thoughts, or if everything I put on paper was somehow impelled by the telepathic contact.

One suddenly interrupts:

> You cannot possibly understand what this means to us. To see— actually see, physically—into another world. …Not yet, and perhaps not at all will you have equipment to see into ours. [Equipment?] We

are after other techniques, other aims, other purposes. We are com-
mitted, you and I, to other ends. After a lifetime of preparation, change
would be...not impossible, but foolish! A great waste! We collaborate
on many thoughts, as we must. We do not want to impose on your vacation and certainly never on your privacy. As you say, "censored!"
We do laugh at that! We are working things out very well. All questions
will be answered in time. All will be revealed possibly, according to
your capacities. They take only refinement, practice now. ...Yes,
guard but do not destroy these notes. Read now, please, in *The Betty
Book.* Many wait...excellent work done this morning. Enough for
now. Enjoy!

—1:10 P.M.

Yes, you are becoming acutely aware of us and our thoughts.
Not all of the chapter you just read is very sensible. "Showing
pictures" like this is really sort of useless. Poor translations. However
much good is suggested therein. We are all learning.... Experiment,
but that is what they were doing, too [those in *The Betty Book*]. Read
on, please. Try not to comment. ...We have our reasons.

"Others present?" I asked.

Exactly.... You have been chosen for more than mere capacities,
though these are many and well defined. But also for your great
spiritual leanings and the sincerity with which you wrote your outline[1].
This is an excellent piece of work, and will be clarified and extended
before long. We are going to work closely with you, but not so much
as to encroach on your privacy or on your time needed for worldly
activities. Your intense cooperation and concentration are needed,
and have been freely given and are extremely appreciated. No
encroachment on your identity or purpose will ever be permitted.
Great powerful ones are watching over this experience, which is
experimental for us all. Many years have been spent in preparation.

...Yes, the writing is a crutch. Soon it will cease and all will be
telepathy. No, we are not "taking you over." This is the last block
between perfecting our technique and getting it all done. You must be
able to distinguish when we are requesting and when you yourself
are deciding. Without conversation and without writing, it is very
difficult, and calls for great awareness. You are getting there.

June 16, 1978—1:30 P.M.

We had gone to Sea Life Park to see the trained dolphins and
whales. We had visited them on other occasions, but it was always a
delight to watch them perform again. *One* had also expressed a particular desire to see the dolphins—he seemed to have a special interest in
them. Before the show, I walked down to the tank and stood nose to
nose with a splendid dolphin—at *One's* request.

---

1. Kannenberg, Ida M. *The Alien Book of Truth.* Newberg, OR: Wild Flower Press, 1993.

...Yes, there was communication with the creatures [dolphins]. Very vital to us. Nothing to you. I, *One*, was with you all the time.

However or wherever you get your ideas is very unimportant. The important thing is they are exactly what we need at the very time we need them. Now we begin to wonder?

Good night now. Many thanks for the dolphins and the reading...*One.*

—2:45 P.M.

We wanted you to know that we are working on the interior physical aspects at the present. Sorry we made you ill momentarily. We were a little over-enthusiastic!

Bill and I had been walking through the marvelous shopping center, Ala Moana, when suddenly I felt a weird, nausea and weakness. I had to lean against the side of a building for a few minutes. Bill, of course, had no idea of what was going on and became quite worried. He tried to recall what we had eaten for lunch, then remembered we had not eaten lunch and went at once to repair this lapse. A glass of Green River and a dish of ice cream at Farrell's cooled my interior and I felt immediately fine again.

We are going to be busy at this for several days. You may feel a bit dizzy or suddenly tired or a bit ill from time to time. Don't worry. It will be a very temporary upset. Only a moment. As for the writing, we will try to have something presentable to show Bill before you go home. Depends on how much you are able to write...*One.*

June 21, 1978—5:00 P.M. (Honolulu)

This day we had tried to see the Iolani Palace, which had been reopened since our last visit, after extensive repairs. We had never seen its interior, and I recalled that one of my unseen correspondents had asked to see it. We discovered, however, that there was a wait of several days. A long list of would-be sightseers was ahead of us. The first day we could see it would be Friday of next week, the day we were going home. So we had to set this aside until our next trip to Hawaii.

—6:00 P.M.

The sight-seeing today was unexpectedly good. Too bad about the palace, but the Mission houses more than made up for it. We saw more of what was along the line of someone's research than the palace probably would have been. ...Yes, if the Bishop Museum is of early Hawaiian people, it may fill in better also. Hope we can do this.

Now! The important thing today was a kind of initiation. Actually we are amazed at our own daring. If we had more time to think about it, we would not have done it! But all went fantastically well. You were superb!

We had gone to the Honolulu Academy of Arts before going to the Mission houses. I had wanted to see the Oriental exhibits, as the main portion of my antique shop was Oriental. This particular exhibit area was closed for redecorating, though a fine little fellow allowed us to take a quick sneak through the darkened and draped rooms to get at least a glimpse of what we could not see. But while we were resting on one of the benches in a courtyard, the inner prompting told me to go back into another room we had just visited and find the statue we had looked at. Vishnu! I was told to look at it steadily and not to take my eyes away or move or even think! This I did to the best of my ability until I was told it was finished, I could go back to my bench or take pictures or whatever. Now came the explanation, in small part, of what had been going on.

> The statue was nothing at all to us, except to use as a concentration point for you and because Vishnu represented to the people of India some of the forces that we wished to call forth today. We still tremble at our own daring. Failure would have been unthinkable! But your splendid cooperation and almost absolute concentration carried us forward more quickly than fifteen years or so (your time) of slow step-by-step experimentation might have been able to do. Our supreme gratitude to you for all of today. We know you cannot do more now. So, Aloha, see you when possible...*One and all.*
>
> Be not forgetful of your mother. Send a card or two. [I phoned.] We take too much of your attention sometimes. Just ignore us now and then! That is all for now...*Victor.*

Victor and I had some altercation over some minor points, but had straightened it all out nicely between us. Some of these fellows had a ribald sense of humor, which I would not permit to take over. I do not really take teasing very well, and sometimes the high spirits of these fellows—whoever they were—carried them away into man-type jokes I did not wholly appreciate.

> Victor is going to be a good fellow now. He is truly mortified. ...Yes, you explained nicely. Thank you...*One.*

—11:10 P.M.
> We are not going to try to explain tonight about Vishnu. We want to, but.... All right. Today you accepted the Order of the Rose. We renamed it especially for you. It is a pouring-in of energies from us. You would not know if we described it to you. You will soon start to detect greater truths underlying surface appearances. Already you do much of this—more than we thought possible for a person of your kind.

I interpolated that maybe this "pouring-in of energies" had occurred during the first pendulum episode, or what I call, "the church incident," that happened back in 1968. They had told me several factions were vying for my attention at that time. Maybe one of them had sent me to that church, where I had certainly concentrated without moving for *six hours!*

Oh my God, Ida, we are so stupid! Of course that is what happened during the church episode. You were given a very large dose of these same energies. No wonder you are able to do some of the things that have baffled even us. Now we must go over our research again and try to determine just who was working with you at that time. Everything was so new and experimental to us at that time, and factions were not synchronized with their efforts. That is why everything got so terribly confused for you and you became ill. We will research this; it should reveal many secrets to us. Thank you for pointing it out.

Now that you have been re-energized once again, new truths will become evident and we will help verify them for you until you become expert. Actually, you are already more expert than we believed possible. That is enough for tonight. See you tomorrow...*One and all.*

June 22, 1978—8:30 A.M.

We shall be very busy all day, working in research and also doing something to your "innards." If you feel suddenly ill and faint, or light-headed and dizzy, don't panic. Just sit down quietly somewhere. We have to rearrange and re-energize some of the molecules and molecular activities. ...Yes, you are right; the whole thing is mostly a matter of molecular health, arrangements and functions. We can do a great deal to energize, reactivate and guide the use of these factors.

Yesterday's episode with Vishnu could have been total disaster (for us, not for you—you would not have known the difference). We staked much on your abilities and willingness to do just what we said, and you did better than we dared hope. Had we remembered the church episode from before, we would not have been so dubious. You were splendid. Our work would have been set back a tremendous amount if we had failed, for we called on forces that could have turned back on ourselves and weakened our presentations. It would have meant many weeks of rebuilding to get back to the point we had reached. All of this sounds mysterious, but there is no way at present we can make it any more clear. It would require too many hours of explanation, and we must go on now with what we have started.

Now we are mixing up our thoughts until neither one of us knows who speaks. Time to stop. Thank you, from...*all and One.*

June 23, 1978—7:20 A.M.

Please give us just a moment. We will be with you today and always. Do give us a thought and a word now and then. But we won't bug you all the time. We realize how much you have given us, but the time has come to give you more freedom.

Go now, have a good time. We all love you. ...Yes, and Bill too. He is a very good fellow and cares more for you than anything. He worries about you a little too much maybe, but only because he loves you. You have taught us what human love can be. I fear our previous idea of love was based mostly on sex. We did not know. We are learning. Forgive us our trespasses! ...Of course you cannot know us now, but you will know all in the very near future. I had not meant to write so much. The pigeons will come back. We love you...*One and all.*

—1:45 P.M. (The Bishop Museum)

We want to thank you for the marvelous entertainment of today. The planetarium show was superb and worth the whole effort we have made so far. We got so much more out of it than that commentator's boring talk about legends. It was like seeing the sky with the eyes of your people, which none of us has ever done.

The museum was interesting to another faction—particularly the thrones and crowns. The whole exhibit was far more valuable to our researchers than the palace could possibly have been. ...Yes, we have something to tell of the original Masonic order, which began in Europe and the Far East centuries ago. And that is why King Kalakaua was important to us, his Masonic connections. Though the Masonic Order of his day (and of yours) is a paltry pip-squeak of an order compared to the original. So much has been distorted and lost! Someday we will go into this further. ...Yes, your episode with Vishnu was a kind of initiation into the "Order of the Rose" renamed, and had connection with the true Masonic background as it was in the beginning. No other woman has been so initiated; that is why we had to rename it. It is forbidden to use the original for a female! We do not hold to that theory, fortunately, or we never would have found you. The Rosicrucians are another distortion and dilution coming from a common source with the Masons. More of that later also. ...Yes, it has become all ceremony and ritual and the real kernel of truth is gone, absolutely gone.

June 25, 1978—3:30 P.M.

Yes, a moment only.... What you are reading *[Unobstructed Universe]*[2] is not a very good explanation of our situation, which is not precisely the same as theirs. They do deal with the subconscious. Our magnetic field is the super-conscious, a different thing altogether.

You are very slightly dissociated—just enough to keep two levels of awareness on stage at all times. You are not, however, using your subconscious, but your super-conscious, for that is our area of activity at the present time. We can and do have another area of activity, which will be shortly brought into your knowledge. It is an important one, for which this preliminary work has been preparing you. *A great surprise is in store for you,* and we hope you will be as thrilled as we are to present it to you.... Patience. A few days only.

2. White, Stewart Edward. *Unobstructed Universe.* New York, NY: E. P. Dutton & Co., 1940.

The statement you truly and accidentally overheard: "If she just knew what is going on here" referred to just that. We simply had too many things going on at once. You were absolutely right to rebel. It is your life, your mind, and in essence, the main part of this whole business is your work, the writing. We are...well not hirelings, my dear, but we should be your obedient servants—only you do not know enough yet to supervise matters yourself. I am, therefore, your surrogate, until you develop enough awareness and knowledge to take a hand in the running of things.

As for the naughty fellows who do not like to shop, they were so amused at your comparison to the men and husbands of your world who hate shopping that they have promised to be good fellows and keep their protests to themselves. Besides, they enjoyed the show. ...Yes, a fringe benefit! [A hula show in the courtyard of King's Alley.]

Victor fed you some balony today. Try not to confuse it with our truths. He is constantly getting his feet in his mouth. Sorry we brought him in...*All and One.*

Suddenly a very authoritative voice interrupted:

Be at peace. We told you to stand up for yourself. These personages take too much advantage of your willingness to be helpful. You gave your word not to let them down. You are released from that vow. Do you want someone new to help collaborate on the books? They will have access to the very same material. You will notice no difference.

Bless you. I am your light. We will continue with "One," the Nameless One. You call him the powerful one. He is sincere and honest. Trust him. He has now been given greater power to control the whole scheme of things. He can now control the others he needs to help.

Then, *One* began to speak:

...Yes, *One* now. We will continue as planned, but on the book only. We will lighten up the daily routine. I know what inspired your speech of protest. Forgive us all. We will do things with reference to your well-being and physical health and with proper respect to your person. ...Yes, your identity. We have been too eager, but that is no excuse.

Thank you for defending me. I would have been heartbroken if I had been taken off this work. We will have Suma and Amorto for co-workers as far as guarding and protecting go. Only you and I will collaborate on books. Thank you, dear...*One.*

June 26, 1978—11:20 A.M.

We have a new audience here—a kind of council. They are debating how to proceed with the writing. Some think one thing and some another. Some want to quit altogether. ...Do what? Vote? I will have to explain to them what that means. ...Yes, wait.

Ida, we never took a vote before—never! It was really hilarious to watch the proceedings, but I did not dare laugh. They were so bewildered that you could talk to us, and they could see you when you went to the mirror and showed them your face....

Oh, this is absolutely true. Believe me, those fellows will never be the same. Chastened is not strong enough word. They were flattened and bewildered.

We are researching our own research! We have been draining you of your energy. We thought we were replacing it by the vibrations and otherwise, but it is not as we believed. Give us time to ascertain just what happens. It may be two or three days before we are ready to write more. ...O.K....sure. You are too nice to tease...*Amorto*.

June 27, 1978—9:36 A.M.

You are so much more advanced mentally than we had expected. And to think we at first called you "the little old lady in tennis shoes!" That was a misnomer! You already know so much, your outline of years ago shows this, and you have given us glimpses of how it was thought out, enabling us to see the good foundation of the careful analysis you have developed. There are high forces who have worked with you all your life, bringing you to this point in time.

But don't fret; you have, by your care, patience and keen understanding, worked out some mighty difficult and obtrusive problems. No one has ever touched on some of these. You have, in essence, welded a religious and scientific outlook. The world needs this now. It is the *right time* for all of this to take place. You will be helped by great powers to come to a point where you will have free time and opportunity to write your books. Now is the time. Personal problems will be resolved without harm or hurt to anyone. Changes—big and important changes—are at hand for you. You will even be happy. (I sneaked that in for you)...*Amorto*.

Your protests were absolutely right and rather vigorously delivered. We understand and love you. When you get home with your splendid typewriter, we will go on with the books. We will contact you going home, but only if the movie is good. Thank you dear...*One*.

Feeling rested and invigorated by my adventures in Hawaii, I soon settled down and began preparing myself for the task that I knew lay ahead of me.

All is Splendid

Three days later, much to my amazement, my friends attempt to explain themselves and their operation to Bill:

June 30, 1978

Ida has been working with us for several months now, but we asked her to keep it very secret until we could devise ways of explaining to her just who we are and what was going on. She is not crazy, and if you were up on what is being done in parapsychology today you would know that others, too, are beginning to contact our area of being, and that a breakthrough is imminent between our various worlds and yours. This coming connection is as important, or more so, than putting a man on the moon.

Ida is unusually adept in receiving and recording our messages, for she has an open mind and neither accepts everything nor rejects anything without a great deal of examination and thought. We can tell more about this later.

Our contact with Ida is solely by telepathy, but we can see through her eyes and hear through her ears at times (but only when she permits us to do so). We cannot invade her mind. We have no way of doing so. All this will be further explained. We are only a small segment of persons in our world able to do this. We could be called the scientists and researchers of our world. Only a very few know anything about what is going on, and while several have worked at various times on this project, only *One* now dictates through telepathy to her (though others will take part at times as other duties call him away). When she wishes us to see something, she contacts us and the particular scientists interested in that exact field are then able to view the subject with her.

Ida is not crazy. We are not figments of her imagination. We are not secondary personalities or thought constructs or anything like that. We are *bona fide* personalities from our own place of being. Ida has been careful not to allow us to impose on anyone in her world—not to allow anything to happen that could embarrass anyone. She has been very outspoken in telling us to mind our own business when we have pressed too closely with our questions or requests. This is enough to introduce ourselves.

Our respects to Ida's husband...*One and All.*

Then they spoke directly to me:

Ida, we are truly sorry for our constant change of plans. We know it is very hard on you to be jerked this way and that, never knowing what we really intend to do. There are reasons we cannot quite control it. None of us is permitted to make final decisions. We must have the approval of other factions. Each faction wants things done in a different way. It causes much confusion.

An unknown entity, calling himself "A Special Person," piped up:

I cannot name myself, but listen carefully. The others will not see this until your book is copied. You have been doing some good analytical work this morning, helped by me. Of course I help you and always have all of your life. Hweig said he did, but he was an interloper. He lied about his part in your life.

I am the one who has directed your life from birth, though I sometimes let others take active part, as I have many duties besides this.

I helped you "guess" this morning. We did come from the same source, you and I, which is entirely hidden to you now. And we did choose different worlds so we might bring them into closer collaboration. This was our chosen task, and we could choose between us who went to which world. I chose this one, rather selfishly for my own reasons, and you took what was left, as you always do! That is why you rebel so much! But you have done splendidly. Free will was a hindrance to you, but you managed to overcome your errors and pick up and go on. We will talk more on this later. This must come to you in small bits, as it will excite you too much all at once. All is well! All is splendid! Be at peace! I know you do not in the least understand what this is all about. Later, you will remember....

One then interjected:

Don't think we are making this whole episode up like a scenario, as you keep saying. *We have to protect ourselves.* We have to figure out how to tell you things. We no more know all and see all than you do. We have to think our way through and experiment, just as you do. We are not frauds. It is only too difficult! Until tomorrow...*One.*

You are right. We are not phonies. We are friends. How to convince you? We cannot. We deplore the seeming discrepancies in our story. They are not really discrepancies, but are different viewpoints of very extensive things.

Do not forget. We all exist in geometric time, and in each area of time our being is different.

I think we must emphasize that again: *We all exist in geometric time, and in each area of time our being is different.* That is perhaps the most important thing we have said to you thus far.

July 8, 1978—5:48 P.M.

After the UFO program tonight, I must tell you something: *The UFOs are our manifestation!* We will tell you more tonight.

Later that evening, *One* made himself known:

We are the originators of what you call UFOs. Some of the UFOs are energy manifestations. We mentioned this once before—or rather how we materialize whatever we need for our own use.

That is, however, not the full story by far. We are just as human as you are! But we are also masters of all psychic phenomena.

Do not believe everything you read about UFOs and their occupants. Especially do not believe the current stories about the airships seen in the late 1800s. Time, in your world, has distorted even more of what were written distortions in the first place. Right now, we must get a vivid outline of ourselves and our places of being into your minds.

To utilize our psychic powers, from our area of being, we have what your psychics call out-of-body experiences, or OBEs. When those of your people have OBEs and have learned how to control them, they, too, will become aware of this coexistent world (coextensive to your own, existing in the same space, but in a different time dimension). It may take many years before they become psychic masters able to do it as we do.

But we do have a very human and very physical existence also, and more than this, we are completely conscious of our being and activities in both at the same time. This sounds incomprehensible if I were to add that we have even more—at least five altogether—areas of awareness of which we are completely conscious of all at the same time. And it would be even more incomprehensible than that to say that each of you [Earth humans] equally has five areas of being. The only difference is, you are aware, or conscious, of only one—the physical one you now are using. This is a very complicated subject.

Back to a subject more to our interest—ourselves! (That is meant humorously.)

When I said we had OBE manifestations or activity from "our area of being," I was referring to our physical area of being which is indeed a planet in a far galaxy.

When we described ourselves as energy existences, we were in no way telling an untruth, though it certainly must have seemed strange to you, as you complained so vociferously. How you do insist on things "making sense!" Many things do not make sense if you do not know the whole story; therefore, be patient and wait for more explanations.

Now we can tell you about ourselves—beings just as human as
yourself, and probably some of the little jokes we dropped here and
there along the way will make sense to you now. I am afraid they only
baffled you at the time. And made you distrust us. However, the time
has come for a more complete exposé of ourselves. Not everything—
just a little at a time.

Have nothing to do with revealing any of this now. If you pick out
various references in your writing, you can just about put a whole story
together. Do not reveal any of this at the present time. Gather your
material, be prepared, and wait.

You wondered why, in Hawaii, we did not want you to go near the
lovely fountain and asked you to hurry by. And why we did not want
you to go closer to the ocean, but asked you to stop and go no further.
As energy existences—and we were with you as such at that time—
we fear the water (particularly moving water) for it can disintegrate our
energy-manifested UFOs. They are sometimes hidden or based un-
derwater. …Yes, even the energy-manifested ones. Remember that
this energy is just on the edge of matter. As the craft rises, the wave
action—not physically, but by the energy generated by the waves—
can destroy the "pattern of being" or "construction" (you would call it)
of the craft, and dissipate the energy patterns of those psychic travel-
ers inside before they could remove themselves. …Yes, I admit that
sounds "awfully odd," as you say, but, here again, wait for another
time and more full explanations. This rising from the waves is a very
dangerous undertaking for us. Remember, just because all this is an
energy manifestation, it does not in *any way* mean it is not real. The
craft and its occupants are in no way figments of imagination. They exist, and the occupants have life and purpose—and they do not want
their psychic existences scrambled, which would take a long period of
recall to recover.

We cannot take time now to tell you more. Put all your notes on
UFOs together, both what we have said and what seems plausible
from your reading, and wait.

Our purpose in entering your world is simply scientific study (just
as you go to the moon), plus the expectation that somehow, sometime, we can communicate and share experiences and knowledge for
the use and betterment of both worlds.

You are invaluable to us, both for information given and for the use
of your vision and hearing. We have been saved much extensive and
difficult research on our own. We do have instruments to see into your
world. We can scan almost anything, and the findings are recorded in
the manner you use film. Also, as energy existences, or in our psychic
being, we can see somewhat dimly into your world—but in no way as
clearly and consistently as you can show us.

...Yes, indeed, we have been using the "confusion technique," as described in your fine new book [*The Invisible College*].[1] That is a valuable book. Study it.

We are all working together to develop your abilities. More exciting revelations will come, but be secretive. Our real work will be revealed later. That is all for now. You are always in full control of your own mind. Think twice... *Your Guardian.*

July 9, 1978—11:20 P.M.

Do not be frightened. This close association of minds is even more dangerous for us than for you. Already you can detect our emotions and, occasionally, thoughts before they are actually transmitted. We cannot have you looking into our world or minds at present. We are using the highest and most refined form of telepathy. We never expected to find an Earth person capable of this. With our careful development and guidance, however, you have attained it.

July 10, 1978—12:30 P.M.

...Fine. ...Yes. The time for frivolity is over. Now serious work. You can listen to people discussing UFOs, and always tell them:

1. They are friendly.
2. They will contact in time. *Soon.*
3. Sometimes they cannot appear in their own form, but must adopt coverings of one kind or another.
4. Crafts are sometimes constructed of metal (or hardware as you call it), and sometimes are psychic manifestations, which then disappear completely.
5. They mean to help Earth people with coming trials.
6. Contactees will increasingly be able to give accurate accounts of the crafts and their occupants.
7. The reason for so much secrecy was their lack of understanding of Earth people.

Please be calm and listen. We do not intend you nor your world any harm. We are who and what we told you.

Of course, we are not miracle workers (as we have said). We cannot wave a wand and by hocus-pocus help you attain all your desires. ...Of course you must follow practical lines of endeavor to obtain any of the things you wish. To grow old gracefully, you must care for your health and appearance. There is no other way. Certainly the right mental attitude is of first importance! To cure illness you must see a medical or other qualified doctor. Certainly there is psychic healing, but of psychosomatic illness mostly. We can direct certain energies into the path of disease or injury and thereby alleviate the condition, or at least the sensation of pain. We cannot cure. Incidentally, psychic surgery is all bunk!

1. Vallée, Jacques. *The Invisible College.* New York, NY: E. P. Dutton & Co., 1975.

Do not confuse the psychic world with the spiritual world, or the
world of departed spirits. This is the place where healing can take
place, miracles can occur. But we have nothing to do with that world.
The psychic world has nothing to do with the spiritual world, but your
world confuses the two absolutely.

There is an immense need to distinguish between psychic and
spiritual worlds. Until tomorrow evening....

July 13, 1978—9:48 A.M.

You ask about the lesser guardian, the guardian of the threshold,
the greater guardian and that sort of thing. This is part of the Rosicrucian teaching and the various stages through which they take their initiates. It has nothing whatsoever to do with us. All of your honors
bestowed—the Order of the Rose, etc.—had nothing behind them
except the desire to confer some nice title on someone who has done
a great deal for us in teaching us the ways of your people on Earth.
We were very much in the dark about many things.

We have had some fierce battles with various factions here as to
how to proceed with you. My word has no more bearing than any other. I am not at all a powerful personage. That was play-acting. I am just
"one of the boys" and I must follow instructions to the letter. Knowing
you better than anyone else at the present time, I can make suggestions that are listened to. In fact, this association has given me more
prestige and notice than I have ever enjoyed. It makes me quite a fellow. You are something of a pet project for many, but to me you are
indeed a pet. I have grown very fond of you. Forget the other; it was
play-acting, a method of testing our capacity to play upon your emotions. From now on, your emotions are your own—so watch the tem-
per. We send out impulsions to arouse emotions in your mind and
then rejoice to find them there! All we have ascertained, really, is that
you are capable of feeling anger, hate, disgust, love, aversion, etc.,
and we knew that before we started. We have wasted time. We must
think... *Victor.*

The Universe Illuminated

From July 10th to July 14th, I was told a final good-bye and a new hello about a dozen times from one faction or another. Over and over they would tell me they had to leave. *"Good-bye,"* they would bellow, and in the next breath they would say they needed my help. Then again, a little later, *"Good-bye, good-bye."* I was beginning to protest loudly, mightily and frequently.

Later they explained their behavior:

> There were indeed a number of separate factions, some of whom bid you good-bye and left, never to return. Some did not go at all. Others tried to leave, then decided they had best stay. Of course it was confusing to you, dear, since you had no way of knowing which faction was which or distinguishing in any way between them.
>
> The major faction that bid a final good-bye at this time were those energy existences (or essences, as they themselves prefer to be called) who had introduced themselves as those who were never born and never died, who neither ate nor slept, and who could never enter into the physical world in any manner, once they had chosen the psychic world. You had no way of distinguishing between these and the psychic or energy existences who were the OBE manifestations of the UFO people. For a long time, such lack of information kept you very confused and frequently angry, for you could only think someone was playing high and mighty with your credulity.

July 14, 1978—10:20 A.M.

> I hate to admit it, but you are correct in your diagnosis. We trap ourselves and reflect ourselves and know your world not at all as you people really are.
>
> We do not know, but are becoming convinced that you, Ida, must be someone special, or at least different. Someone sent to guide us or save us from making complete asses of ourselves, and doing harm, quite unintentionally, to your people.
>
> We are changing. You will read now of changed or different types of experiences as we have contact with your people...*Mordalla.*

Well, here was a new one—"Mordalla." He continued:

> Dear girl, you are so right once again. We did forget about the rats-in-the-maze experiment. They did become psychotic when confused over endless periods of time. So we will stop this confusion. We do have a purpose for coming into your world—and it is a good purpose—for both our worlds. We are not trying to control people's minds—only to awaken them and make them think. We want Earth people to know about telepathy, hypnosis and dreams—all psychic, not spiritual phenomena. We try to give them experiences so they can analyze and discover truths for themselves. Perhaps this is too much to expect.

> We can only dictate, then, and with your help, get these things published. Yes, we try to get all of our contacts to write of their experiences so truths may be ascertained from the overall pattern. We had no idea people's minds were so disconnected from each other. You do not work together and share information as we do. You do not truly cooperate, although you are fumbling in that direction.

> We will continue at once putting your pendulum story together. We will not dictate so much on this as collaborate, but you must choose your manner of writing. We are not very good at it. Your everyday speech is different from your manner of writing, another thing we did not consider. After we see how you work out some of the writing, we can help there and save some of the time we have so stupidly lost. We are arrogant and egotistical; you are right to chastise us. We do not want mirror reflections of ourselves. We, too, believe tigers and zebras and giraffes are far more exciting than looking in a mirror! How well you point out these things. Why should we go so far only to reflect ourselves. We are indeed from "outer space," as you call it, and no one yet has been told the truth of exactly where. That is all we are prepared to say at this time.

> We are able to see through your eyes and hear through your ears. We did see all the wonderful things you tried to show us, and it was a very pleasant and worthwhile experience. Some of us had seen something of all this before; some had never seen any of this. Seeing it through your eyes made it so much more real, instead of a picture on a "film" type of thing, which is how we see it through our mechanical reports. It is somewhat analogous to your TV, but actually not at all the same. ...No, we have nothing to do with the Masonic Order or the Rosicrucians, but we do know about the original practices and secrets.

> Now you may not accept this, but the lost records of Atlantis were very real to us and you did have the key to their whereabouts—but why it was given to you or by whom, we will never know. We have worked and researched and talked and many have spent considerable time trying to analyze why you were given such a secret.

> We are persons quite a lot like yourself. More about that later. There are other varieties of us, some of whom would seriously frighten you with their visage, but interiorly, they are as human as you and I.

Some of those who appear to your people are psychic manifestations—that is, illusions. Do not confuse psychic with spiritual. We are
going to say this over and over again.

In the first place, we do not want to make people alike to us, or to
control them or make puppets of them. What would we do with them?

We thought we could help your people evolve, but we see this is
absolutely the wrong way to go about it. The only useful way to help
humans is to help them find the knowledge and develop their natural
skills and capacities and capabilities so they, too, may use the same
powers inherent in themselves, which have fallen into disuse. Your
kind of proposed writing can help. We have ceased trying to hasten a
process that must come slowly, through step-by-step acquisition of
knowledge and practice. Therefore, we are withdrawing such intervention in men and women's minds and growth, as has been exhibited in many places recently. Instead, people will be inspired to
experiment, learn and write, and these topics will become common
knowledge. You have your story to do first, and your experiences evidence the conflicts in both worlds, for our conflicts are many.

This is something you must keep in mind as we go forward now—
that we are not free agents (not as much as you are) and we have
many factions here that we must keep appeased as we go along. Not
all are in favor of continuing with you. Many are ready to stop at this.
But we few feel you have too many capabilities to let them go to waste.
We truly need your help.

Go, now, to your work. Later we may discuss this more thoroughly...*Mordalla.*

July 15, 1978—3:15 A.M.

I had been awakened by a strong electrical storm. I went to the
bathroom for a drink of water. A great flash of lightning seemed to
strike along the back fence, just as I reached for the faucet. I gave up
all idea of getting a drink and scurried back to bed.

We have come to awaken, to make aware, to make you determine to do something. The task will be revealed later. We have vastly
superior technology and vast psychic powers, but we are the drudges
of the universe. We are minions. We come to prepare you for great
events to come. You can turn these events for good or evil. *You have
free will.*

Now what in the world did all of that mean? The voice continued:

One of the purposes of this account is to demonstrate one fact
through experienced events and revealed principles. That one fact is
that psyche—or soul—and spirit are in no way synonymous. You will
not agree with this immediately, but ultimately, you will.

There are two divergent modes of being that actually have nothing
in common, save they happen or belong to each single individual being. Each individual has—or is—a soul, or psychic-being, and each
has a spiritual reality quite separate. The "psychic world" and the "spir-

itual world" are separate. Confusing soul and spirit is quite easy. Un-
confusing these separate areas of being is not so easy, as generation
after generation have made no distinction between them.

Coming events make it imperative that the difference, and the
separated sphere of each, be properly analyzed and revealed.

July 16, 1978—11:15 P.M.

A resumé for your clearer understanding: We are the "UFO people," as you refer to us. We do come from a far planet, though we have
some permanent bases on Earth, beneath the sea and otherwise.
Some of our ships are "hardware" and their occupants are human beings very much like yourself, only far advanced in technology and in
psychic abilities. Many of our ships are psychic creations and their occupants are perceived incorrectly due to hypnosis of the contactee.

Our planet is full of color and has gorgeous flowers, but we have
no animals or birds. They were all destroyed eons ago, except for
those brought in and kept under scientific conditions. These are many
and varied, and from many sources.

We travel through a time differential. I am not at liberty to explain
nor to reveal our propulsion system, except to say it is something akin
to anti-magnetic forces.

Our purpose is friendly and for the mutual benefit of both planets,
and other planets and worlds, for we are of a confederation of planets.
Great events are coming that you must be prepared to meet, and we
are here to help you become alert, to prepare and to understand what
is happening. It is *vital* to our welfare that you are successful. If you go
down the tubes, we and many more go with you.

We have made many errors of judgment in dealing with Earth
people, for we did not know enough about their thoughts, feelings,
fears and desires. Such contacts, as we have had with people of your
Earth, have given us information along other lines. Most of those who
could converse with us gave technical or mechanical information. We
had not learned the everyday thoughts and feelings and attributes of
your kind.

Because you are so primitive technologically and so undeveloped
psychically, we greatly underrated your capacities of understanding
and learning. We thought of you as experiments and something to be
reconstructed according to our blueprints. Now we know we have
only to explain and teach and help you develop yourselves along the
necessary lines.

First we must gain the confidence of your people—and, up to now,
we have done a poor job of this. Meddling with your minds is not the
way to do it.

Rather we intend to teach you how to avoid mind control by anyone. You and others like you will write books and teach all these things
you must learn.

Not one of you has any mission to save the world. All of you working in good faith with each other, with good will and a common purpose, can accomplish what is necessary in the proper time.

Not one of you is more important than any other. One thing to avoid at all costs, is to let any kind of cult or religion grow out of what you are taught.

You, personally, shall not endeavor to help others with abilities you may develop, for this would only put a stop to your real work, which is writing, collaboration, and inspiration.

Earth people have two questions concerning us that we cannot account for. The strange Men-In-Black and the animal mutilations are not our doing! We are going to make a strong effort to find the answer to these riddles.

At present, only one planet is constructing and sending ships, but with the help of several other planetary specialists. They are involved in making plans and decisions with us.

We are beings like yourself for the most part, though there are a variety of what you call "humanoids," who are quite different. They are mostly workers and those who function best in routine and uncreative jobs. Persons like myself and you are the scientists and planners and supervisors. Each of us is scientifically chosen for his or her appointed task.

As persons, we live a long, long time, so long many forget there is such a thing as birth or death. These are natural occurrences here, just as they are there. We do not marry and have families. Children are born to a few select women by a few select men. Your concepts of marriage, family and home are quite alien to us.

You wonder why we can converse so easily in your own kind of speech. Our language is mostly pictorial telepathy with each other, but there is a person who is able to translate for us and is acting as interpreter for all of our messages. We transmit through Mordalla. It is translated and relayed on your side by one who calls himself "Victor."

We have asked you several times to please wear black and white clothing. This is because of the factors of visual reproduction. Colors become blurred and off-color in transmission. Black and white keeps a sharp, clear outline, and it "comes into focus" better.

You have protested that much of what we tell you sounds like feedback from things you have told or shown or read for us. There is a good reason for this—the transmission of ideas by direct verbal telepathy. We can only use the words and visual experiences you have in your mind already for this kind of contact. If we tried to explain something totally new to you, we would have to use pictorial telepathy, which takes a lot of preparation. We have given you some exercises in this, as you complained at 3:00 A.M. one morning, but you have not had time to work at this enough to make quick transmission or reception very feasible. Right now time is important. ...No, not even Victor can translate, except in terms you already recognize.

At this time, tell no one about this, except your husband. You have never kept secrets from him, and we do not want you to do so on our account. We do not want to cause strain or friction between you in any way.

Just quietly continue your writing and do not worry about anything. All your personal problems will work out satisfactorily. We can impel some changes in a minor way, through other channels we cannot reveal at this time.

We cannot promise you great rewards—only the personal satisfaction you get in doing something useful and creative.

We repeat: *We are not miracle workers.* We do not control spiritual matters. These are outside our sphere of activity as much as they are outside your own. We pray just as hard as you do for justice and mercy and peace.

Nowhere will you find a completely truthful account of any of our contacts with Earth people. Always we put limitations on the conscious memory of those we contact, and frequently what they think they can see is a hypnotic illusion. Couple that with the fact that much of our manifestations are psychic, and it is impossible for anyone to relate the entire truth about any contact. Our reason for this subterfuge is simple: We are not ready to blow our cover yet—but the time is coming soon.

As we become more adept at dealing with your people (and as they lose their panic and fear, and we are able to deal with them on a more normal basis), we can then make our contacts more matter-of-fact and entirely real. We see this as something we must strive for intently, because as long as we frighten and create illusion and distorted memories, we cannot expect to be accepted as we are. We have wasted much time trying to do impossible things. Now we are working frantically on a new modus operandi.

Our intentions will be suspect until we are able to demonstrate in a practical and conclusive manner that we mean no harm. Rays, beams and hypnosis certainly are not going to accomplish this, and, in fact, will work against our desire to come as friends.

None of our scientists expect to come into personal contact with any of you. This is for the navigators and the specially trained personnel. We scientists must stay at home with our instruments and gadgets and special abilities, while others go traversing through the time differential (and sometimes through space as well), to converse face-to-face with the Earth people. Each of us has particular work, and we never go outside its confines and limitations. Something our leaders do not want us to learn too much about is the freedom with which you can change jobs and travel where you will. Sometime I will be able to tell you about our levels of work, etc. Our social form is not at all communistic, for we have absolutely free choice, up to a certain point of development, as to what we want to do. Each step upward reduces our potential choices, until, after a certain level, we make one final choice and accept its restrictions. We have free will in all things. But after this final choice, there is nowhere else to go, unless we choose to give up everything we have worked for hundreds of years and start all over again—which means more hundreds of years to reach any kind of elevation again. By that time, no one wants to change! Our advancement is more like an initiation. There is no going back any more than you can unlearn your multiplication tables. For us to change jobs as you do would shake up the whole structure. It takes too many hundreds of years to reach the peak skill and knowledge. Even then, we could if we wanted to. It is just never done.

I am trying to anticipate some of the many questions I know you want to ask.

Since our conversations with each other are almost wholly by pictorial telepathy, we have symbols to "flash" rather than spoken names.

Picture telepathy is far different from the picture writing of primitive people or even from that of advanced civilizations like the Egyptians of your world. Our pictures are moving pictures. They show action and intent.

Sometimes it is necessary to do things in a mysterious way just to arouse interest and make people want to know more and to take an active part. For this reason, we did a lot of misleading right in the beginning. We could have told you—and others—a great deal more right from the start, but we felt you would all lose interest if you felt you were being asked to do rather prosaic tasks for rather flat rewards. So we tried to keep you interested by being changeable. I guess we changed and changed to the point of exhaustion for you.

Also, we know now that we went overboard with our compliments, which you took for flattery and became annoyed. Again, we offered too many rewards, which we could not possibly fulfill and which only held us up to inspection as phonies. All in all, our psychological approach would not have satisfied a moron, which you certainly are not.

So sorry to put you through so much. You are now fairly well prepared for your task. Do the best you can. It will not be perfect, but it is necessary to get it done in some fashion as soon as possible.

We have been careful to create no psychic manifestations around you or in your larger areas, so as not to attract attention toward you in any way. When your first writing chores are nearly done, you will gradually find yourself capable of some psychic abilities. Do not panic. They will not be great, just pleasant, little fun things—until we contact you again and we go into more serious experiments together.

Now that we know so much more about your people, we can save great effort and time in getting our very necessary cooperative efforts underway. Others will be helping and advancing our mutual cause at the same time. Many are involved here—more than you can imagine—and gradually many others will become involved there on Earth.

…No, your first pendulum episode will not frighten people away. In the kind of minds we need, it will only arouse a longing to discover the meaning of it all. The second episode will create desire to have similar experiences. You have discharged your responsibility by telling of the dangers involved and the lack of accomplishment, so far as you were able to tell.

Be sure to use all our written material, no matter how many times it seemingly contradicts itself. This is part of our plan, and serves a very beneficial purpose to you in the way of development along several lines. You will discover these things as you write and become free to analyze and interpret. You are still being held back, which is necessary but temporary.

You are far more psychically developed after all these events of the last six weeks than you can possibly realize. Gradually everything will be clear to you, one thing after another.

A little after midnight, the message ceased abruptly. My calls for Mordalla were not answered. I offered a prayer for the good works of all and good care for friends.

July 17, 1978—9:35 P.M.

The show upset you. It seems so unreal! They did see psychic manifestation, not real flesh-and-blood creatures. But please do not discuss this yet. We are not ready!

This was in reference to a TV show about UFOs that we all had watched. Little green men or creatures in green envelopes had almost attacked an isolated house. They burbled strange noises and a man in the house shot at them.

Please work on the pendulum story. You will know how to use the material as you go along…. Very important…*Love, Mordalla.*
…Fine. You will not have trouble remembering my name again.

—10:30 P.M.

Just a little reminder. Do not discuss psychic things or UFOs at all for the present. You may unintentionally say something we do not want exposed yet.

You are O.K. Don't worry about your head. You will hear no more echoes of our thoughts. You will have some directives, but only to keep you from becoming confused or being afraid. We realize how impossible all this seems. Actually we find it hard to believe we have found someone so adaptable to our purposes.

Human beings have been influenced interiorly since the beginning of time without realizing it or knowing its source. You know what is happening, and that is the only difference. We were just as much in contact the last seventeen years, but you were not ready to know it. We first tried direct contact ten years ago, but others intervened and drove us out. Now we have refined our approach and found new helpers to interpret and intercede for us. We do not come to you directly, but through good friends, of whom Victor is one.

You were first picked up on our research instruments (as all our contacts are), then analyzed, followed with our recording devices, and impelled (with the aid of the energy existences), to use the pendulum. Many other people are being surveyed and coaxed by one means or another to make or find contact with us.

Some things our machines do not register; inner emotions and certain states of mind are hidden. There are some areas we are not permitted to scan without the individual's permission, and some are inviolate, just as we can never invade certain areas of your privacy.

A great number of our scientists here are working with persons on Earth to try to bring them to a condition of usefulness for the purpose of making direct contact and physical landings. Your people are not as emotionally prepared for us as we had hoped.

Therefore, we are very eager for you to write of your experiences. The first part of your book will be a warning of the tribulations of someone trying to understand what is happening (as we make many tests to see if that one is to be trusted and to be useful). Hereafter, persons will find it easier, as we now know how better to handle things (thanks in great part to what you have told us and allowed us to see of the thoughts, feelings, and home life of Earth people).

We have come to the conclusion that the only way we can make direct contact without being shot down is by revealing truths about ourselves. By doing it in this manner, we still are not putting ourselves in any jeopardy. Do not talk on these subjects now or reveal anything, except to your husband (if you have to!).

We understand that you do not like—or you fear—the double talk in your mind, so it shall cease for the most part. If you want to tell us something, it will be recorded. It is not necessary to articulate; we are skilled enough now that it is not necessary (at first it was). This is just as new to us as it is to you. No one has been able to do this for us before. You have saved us years and years of fumbling and trial and error experiment. Now we can progress much faster and your people will not be subjected to such fear-provoking experiences. The last thing we want to do is make you afraid of us! We see that it will take much reassurance to alleviate the fear and distress that we have already caused.

Remember, you will be guided and inspired. Greater ones than we have promised you that, and have validated *One* and Mordalla/ Victor as the ones to supervise the carrying out of these projects. I, Mordalla, am glad you feel friendly toward me after all the nonsense we found it essential to perpetrate for various reasons.

I do laugh a lot, but, as you say, some of it was forced just to keep you from getting depressed or fearful. It will only be honest laughter from now on, and you will seldom notice it (or the sighs). Gradually, you will not remember about them. We have learned that we have to do things very gently and gradually. A lot of our error has been through haste or lack of consideration for the person. We are learning what fine creatures you are, and that, as you say, "The mind is a subtle and delicate thing."

Of course, we are as physical as you are, and of course, we have brains, but we are accustomed to using them in different modes of communication and even thought. Therefore, we were not prepared to handle yours with any finesse. We were rather brutal.

You have revealed so much to us that dozens of our scientists are working constantly to analyze and understand the principles of it all. Your lifestyle is quite different from ours.

Yes, as physical beings, we do eat and sleep. The energy existences do not. Then, of course, the psychic projections are temporary and have no valid life of any kind. There are several varieties of physical existences here also—human, humanoids not of the same build or nervous systems as yourself. So you have a great variety of "aliens" who might be sighted by your contactees.

There is so much we cannot reveal until we are sure of our reception among your people. We do not want anything to go amiss when we have made such fantastic progress in recent months.

We are not allowed to initiate, instigate or create thoughts, ideas or beliefs in your mind. But once these come up in your talk, reading, TV or self-analysis, we are permitted to verify, strengthen and, to a small degree, extend the idea or thought. This is one of the reasons why you complain that what we tell you seems to be only an echo of something you have known elsewhere. We cannot initiate new ideas, for that would be tampering with your beliefs. When we telepathically communicate, we cannot do this. (If we could converse face to face, we could tell you anything.)

—11:40 P.M.

Our good friend Victor continues to translate for us. We appreciate the help you are giving us in handling other communicants [contactees]. It has already helped enormously to allay their fears and to make them accept us for what we are.

We had not realized how much fear and doubt was holding them back. We cannot read their emotions, interiorly, until they are expressed in physical action or in speech. Even silent speech or articulated thought tell us something. Emotions that are not verbalized are not transmitted.

We want, at all times, to emphasize the difference between psychic and spiritual. We never dreamed there could be so much confusion. We do not want cults or religions started because of us. That phrase you read, "Chariots of the Gods," startled us immeasurably. We are in no way superhuman. We do have tremendous psychic abilities and we want to help your people develop these same abilities. Once they had these powers, but they have been forgotten. Religious leaders took over the usage of these powers and made secret all the laws and principles thereof, denying the common people the right to develop them. They were lost to the majority of the people. The potential is still there; the psychic centers only need to be awakened and cultivated properly. From the beginning, we want to make these lost secrets available to your world—just as you have made a lost secret available to us. (We are forever in your debt. We simply cannot understand why you were given this secret about the Atlantis records or by whom. It is a great puzzle to us all.)

Never use the pendulum again. This creates self-hypnosis, which leads to dissociation and allows various influences to take over, sometimes to a disastrous degree. You are O.K. now; there is no danger and we can all breathe easier. ...No, goose, no one is locked up inside your head—just your own dear little brain! [This was another word from Victor.]

You will gradually develop some psychic powers, but we don't yet want you to do anything to draw public attention.... Time has run out for tonight. See you tomorrow.

...No, we do not have wild animals, birds or fish. We eat fruit, vegetables and synthetic foods. The sight of your meat foodstuffs disgusts us. Sorry. Like cannibals! Love from us all...*Mordalla.*

Tools of the Trade

July 18, 1978—9:48 P.M.

We are certain you will be discreet and not reveal too much before we are prepared. ...True, people won't believe you out of context. Therefore, we want to explain how the pendulum works.

You speak aloud and the vibrations of your voice, which are a kind of energy, are picked up by what we shall call "psychic antennae." Of course this is only a label that doesn't tell you anything about what it actually is, but for right now we won't go into it any more explicitly. We are tuned in by means of our psychic receivers. You have these as well, but don't know how to use them. You will learn. The vibrations of your voice are at first necessary. Later you learn, through practice, to transmit thought vibrations without actually realizing you are doing it, or how you do it. The vibrations of your spoken voice (or thoughts) received by us are then translated into symbols. We understand these symbols, regardless of your language or any incoherence of speech (i.e., speech impediment, stuttering, stammering, drunken babble, etc.), for it is actually the idea or thought behind the sound vibrations that comes through. At first it needs sound vibration. Later, it can stand alone as thought vibrations, which are finer, more subtle stuff—a different kind of energy. The translation into symbols of meaning are recorded by a kind of chart or graph, from which we read your intended communication. Later, when thought vibrations come through alone and we have learned your particular use of words (for everyone puts different connotations on the same words), we can receive your thought vibrations directly into our minds, as you are transmitting directly from your mind. Those are basically important data on your sending and our receiving.

Our initial attempts at sending to you, before you use the pendulums, Ouija board, hypnosis or any other means of contact, are simply to send a shower of impulses (or energy waves), directed specifically to you. You feel it then as a curiosity and an urge to try to make contact with "whatever is out there." Whenever we note even a spark of interest, we pile on the impulses until the person feels compelled to work further.

Once he or she takes up something like the pendulum, we try to implant suggestions for better use, but we cannot do this to the extent of controlling his actions. He must always retain free will and his mind must never be tampered with. Sometimes we have, in our eagerness to develop a likely communicant, overstepped our authority and only made matters worse by our enthusiasm. We have frightened people away, or caused them much mental confusion, even anguish. We are experimenting always on better ways of doing these things. We try to be easy and fair and not influence the person beyond the degree of his desires to communicate with someone or something "out there."

After we have impelled him to use the pendulum, we must move the pendulum, of course. This is done by direct impulses—or waves of energies—on the pendulum itself! It has nothing to do with subconsciousness or involuntary muscular action. Any jerking of the muscles is due to muscular tension (letting go, i.e., relaxing in one jump). Fatigue, excitement and other conditions cause the muscles to tighten up so that the pendulum cannot swing freely. Too tight a pinch on the string can hamper action.

You had evidence of this direct energy moving the pendulum when you saw the "blue flame" around your little silver pendulum. Too much energy concentrated for too long at a time can become apparent as a light, intensely blue and vivid.

Persons usually get the idea of "yes" and "no" very quickly. They make lines of some kind to designate these responses, write the words on a piece of paper, or otherwise suggest a code for negative and affirmative, which we then adopt. We try by many means to suggest somehow to them that they outline an alphabet so we may clearly make statements and answer or ask questions.

While this is going on, we are constantly bombarding their brains with electrical impulses. Psychic energy is a subtle and refined form of electromagnetic forces. The mind is a magnetic field. The "electric" impulses we send become electromagnetic forces, once they can reach and react with the brain. All your psychic abilities originate in your own brain—the centers are there— but someone must send electrical impulses of this kind to awaken and develop the psychic centers. We are by no means the only entities capable of doing this. Our ratio is very small of all psychic awakenings. But each entity sending impulses does it in precisely the same manner. Some of the others do not have the mental and moral laws and restrictions that govern us; therefore, they are capable of doing much mischief or even unintentional damage. Sometimes they interfere with our transmission of impulses. If they are strong enough—or if several concentrate together (they usually cannot agree well enough for that)—they can throw our work out of gear and cause great distress to the person receiving. They can imitate us and garble our messages and do a great deal of harm to the work we are trying to do, as well as harm to the person receiving. That is what happened to you the first time.

—11:04 P.M.

We are in the process of revising our methods of dealing with persons who we wish to attract to psychic phenomena. Already we have had great success with those just starting—or on the verge of starting—due to those things we have only recently learned.

The pendulum is but the open gate to automatic writing and/or direct telepathy. Some persons do not have the capacity for these, either one or both; some can do them quite easily.

The person watching the pendulum is engaged in an act of self-hypnosis, which opens the mind for telepathy. It is a partially dissociated state. Once the door is open, we try not to let it shut again or let others find their way in to use it. This means we must monitor the communication constantly and therefore limit the number we can work with at any one time. Some are too fearful or have minds that are too closed for these methods to be used. In these cases, we try other methods of psychic contact.

We definitely choose those we wish to communicate with; they do not choose us specifically. Usually, because of ignorance, they are willing to accept whoever answers their call. They can be led into much travail by allowing this to happen.

We choose those we wish to reach by very scientific analysis. They are watched and analyzed for a long time. They are in no way interfered with, because interference would render the analysis futile.

When we have selected one we wish to reach, we send out electrical impulses that are, in effect, thought forms—i.e., not ideas, but pictorial representations very much like TV. (It is not mechanical however, but pictorial telepathy!) Since their psychic centers are not yet awakened, they do not receive these as pictures, but as urges to do certain things (such as use a pendulum or Ouija board, seek a hypnotist, etc.).

We are beginning to understand more and more how to deal—or, more importantly, how *not* to deal—with those we wish to make communicants.

Pictorial telepathy is used until the persons have advanced to the point that they can receive verbal telepathy. This usually upsets them to no end. Usually, however, automatic writing comes first and we are able to warn them that they will hear people talking, or something of that nature. Whatever the order of progression is, sooner or later we are conversing in their own vernacular and idiom, by pen or voice or both. By that time, we have been able to measure very accurately their use to us, to themselves and to fellow beings. We have a pretty good understanding of their potentials as psychic researchers and a very good judgment of their character.

Altogether, our requirements are very strict. Only a small portion of neophytes ever become initiates and only a handful become adepts. It is very rare for one to become a master, even here where we have practiced psychic disciplines for many centuries, and have lifespans of a few centuries. No new masters now come into being; we were all formed long ago, and the advance to master just no longer happens.

Our development is comparatively much slower than in your world. Procrastination is our worst character fault. We always feel that we have a hundred more years to get a job done. We do not "die," we simply retire into other areas of being. We are born, but so long ago we have forgotten. Childhood has no meaning for us. We do not see the few children born here until they are young adults.

Here, *you* are the alien life form! Therefore, we have made many errors of judgment we would not have made had we been able to establish better communication earlier.

The pendulum is one of the best initial contacts because it works so quickly. The Ouija board planchette takes a lot more energy to move, and many people cannot use it to best advantage. Sometimes it takes two because of certain controls that multiply with two. (I cannot explain that now. We have too many other things to consider.)

We are considering making use of several new methods of contact, so as to multiply the communicants as rapidly as possible. With our new knowledge, we can handle more and develop them more quickly!

You have complained of a feeling that someone is moving "all the furniture around in your forehead"—now here, now there. This feeling is the result of our electrical impulses received by the brain. Certain psychic centers are thus awakened, developed or strengthened. The feeling will go on for some time—not consistently, but now and then. We also work on mind components such as memory, imagination and desire. These all have certain brain areas to be vitalized and clarified. Your memory should improve remarkably. [It has.] The old phrenologists were not too wrong by theory, but quite wrong in the way they elucidated that theory. There are brain areas for certain capacities. These must be strengthened and even "cleaned out." Certain old beliefs must go. Certain misconceptions result in stresses and blocks—actual physical manifestations that can be physically removed. Yes, one night, in the middle of the night, you complained that someone was "dumping your brain out." We were indeed emptying some sub-conscious garbage that was only holding up progress.

The energy and the physical are much more closely connected than you realize. Yes, that is a fair analogy, like two sides of a coin. Good enough for now but to be greatly clarified later. You have several times been frightened by strange sensations in your brain, such as the one when we closed the gap of dissociation, an actual physical, but very tiny, gap. You said it felt like we were trying to flip your brain over like a pancake, and you fought against it. That is what you were supposed to do. The pressure you gave was the physical thrust necessary to close it. We cannot move you physically, touch you or make any actual physical contact. This is not because we are unable in any way, but because it is a rule we must obey if we are to work with you in any manner. We have stringent laws and swift and inescapable punishment. No one dares disobey. Our laws are as exacting as the Japanese honor code of hara-kiri, though not of that nature.

You sometimes complain of vibrations in your arm, leg, abdomen, chest, etc. These are electrical impulses aimed as strongly as possible at some disease or ailment we are trying to alleviate. We are only about fifty percent successful, but are working to improve our methods and results. We need help from your area on this, from a doctor perhaps or someone knowledgeable in human anatomy. Dear heart, you don't even know where your gall bladder is!

We mentioned physical blocks and stresses, resulting from wrong beliefs or misconceptions that must be removed physically. We cannot physically manipulate you in any way, so we direct our electrical impulses to these blocks and stresses with the result that they come into your mind, appearing as misconceptions or wrong beliefs, but with a great uneasiness at the same time. You are thus conditioned to work on them as mental problems to solve. We continue to lay on the impulses until you work them through, solving the problems and removing the physical stresses or blocks. The energy and the physical are so closely connected here that I cannot think of an analogy to clarify. The interchange is swift and constant, like lightning flashes darting furiously back and forth between two clouds.

Now as to the manner in which we see through your eyes and hear through your ears: The pulsations set up by our energy contacts with the human brain allow specific areas of the brain to send out or transmit whatever focus of attention this area serves. If it is the visual area, a visual report or transmission is sent out, automatically and unknowingly. If it is the auditory area, a transmission of sound is sent out. (Remember both vision and hearing depend on physical vibrations.) Here again, the energy and the physical aspects interact to produce the results. The transmission is guided by the scope and intent of our energy contact, and can be captured by very sophisticated machinery, received directly by a mind here, or both simultaneously. We could state this more precisely by scientific formulas and jargon, but that is exactly what we do not wish to do. We wish to state it simply enough that anyone can, with a little study, discover its meaning.

This is enough for one day. We are beginning to think with each other's heads, and that makes me just as uneasy as it does you.

Don't worry about anything. All will be fine. Love from all...*Mordalla.*

Art of Continued Existence

July 19, 1978—3:00 P.M.

Although we are, in both worlds, of human construction, we do have evolutionary differences. Therefore, we have emotional reactions that differ, differences in ideas and beliefs, differences in actions and thoughts.

We have been doing research in past records of pendulum users. We find that, with the new keys you have given us, we can analyze most of the formerly mystifying actions and statements of those users.

We are now going to proceed with plans that have been long in the making—certain kinds of contacts not related to UFOs. Why send machines so far if psychic manifestations can accomplish the same thing? With all of our new information, we can at least start trying to pick out certain persons for certain projects.

Do not be so concerned that you may be giving vital information to the wrong people. There is no way at present we can validate ourselves, but we hope some events will soon come to pass—events that we have been trying to engineer—that will bolster your confidence in us. We cannot interfere in Earth events; we can only put a little pressure on something to impel it to happen, once a start has been made. Our disadvantages are many: distance, lack of complete understanding of all factors involved, etc. We were just as baffled as you as to the meaning and purpose of your responses and reactions and the responses and reactions of other communicants. Sometimes you scared us to death!

—9:55 P.M.

Dear girl, now we are beginning to make beautiful sense of our research with Earth persons. None of it made much sense before. We got the effect of a mad house! Everything was chopped up into pieces. There seems to be no coherence to people and their lives there. Every year or two they change their places of living, their job, or even their spouses and children! Chop, chop, chop. Everything done in pieces. Now we have seen how they are geared to this existence. Everyone tries to do and be everything to all people. What a terrible strain! We have watched you nearly torn apart by the various things you have to do.

—11:15 P.M.

We can begin to explain ourselves now.

We could not tell you the truth about ourselves too much for fear you would qualify your answers to our questions according to what you thought we wanted to hear. You are so considerate of your listeners that you mold or bend your responses to their needs or moods and, in most cases, of social intercourse, this is a splendid and commendable course. More people should develop such a habit.

However, in our case, we had to have exact answers regardless of our feelings. When you thought you had hurt our feelings, or apologized if you had made us angry, we were delighted, for it only proved you had answered us directly and honestly.

July 21, 1978

This day I went to my daughter's town, a hundred miles away, to see the two granddaughters, ages 15 and 16, show their goats and chickens in the annual 4-H Fair. Of course, my unseen guests tagged along.

Today gave us a beautiful glimpse into the lives of children and their rapport with animals. The evidence of good sportsmanship among the children was very heartening. We were beginning to have a rather dim view of Earth persons in some instances. Your children have indeed redeemed your people completely in our eyes!

It is too bad there was not more time, and you had to do everything at such a rapid pace. We would have liked to examine the animal barns and the animals more closely, but this quick view was exciting enough. It is all recorded for detailed study by students who have read about such creatures, but never expected to see any.

You have asked about the physical aspects of our world. Our atmosphere is somewhat different. Many of the humanoids would have to wear masks in your world until their lungs adjusted. The humans such as yourself could get along quite well if they did not have to exert themselves until they became acclimated.

Now we must expose this: We alternate between two fairly close planets in a far galaxy. One is our work planet. The atmosphere there is quite bad. It has little sunshine; heavy low, black clouds; constant thunderstorms with torrential rain; and no snow or natural ice. Its climate is fairly temperate, without much variation. It is muggy; even the rain and occasional violent winds are warm. You see, we are able to regulate our temperature and climate through various natural controls. You will learn to do this in the future.

Now the second planet is our residence or home planet. It is quite different. It is twice the size of Earth. We cannot be more explicit now.

We are as human as you are—no more, no less—but we are far more advanced technologically and scientifically, and are masters of psychic phenomena.

We have communication and transport to several other planets, not nearly as far from us as yourself. Earth is the farthest from us that we have travelled in actual physical ships. We travel through a time differential to get there for the most part, and do only a small amount of traveling in space. More on that later.

Hundreds of years ago, we destroyed all animal and nearly all human life on this planet through our activities with atomic and other powers you have not heard of. We hope you never discover some of these! The only animals left are those brought in from other places and kept in scientific compounds. I emphasize our animal life, as I see you are so concerned about the animals of your planet.

Only a handful of humans were left, after our near self-annihilation, to repopulate the planet. Fortunately, great knowledge was contained in these few hundred people and from that time we have been rebuilding, aided by other worlds whose occupants have come to our aid.

We know the people of Earth have reached a point at which they can do the same act of self-destruction that we did. This does *not* have to happen, but it can, and the probabilities are enormous that it will, unless more *knowledge* is quickly forthcoming.

That is where we come in. We want to keep you from blowing yourselves out of the sky. Any great upset in planetary position would simply upset the equalized activity of the universe, including us and our helper planets. We cannot stand idly by and watch this happen. We dare not. We must take a hand. We want to do it in a cooperative way without controlling, but if the people of planet Earth do not listen and take advice, we may have to take compelling action, just to save ourselves, and you! You can see how logical our stand is, and how necessary interference might become.

We are spiritual beings in the same manner and degree you are (although you do not know your own true spiritual reality yet). Therefore, no one must take us to be "gods from outer space." That is pure nonsense and dangerous. We do not want any cults or religions blossoming because of us. That would be disastrous to sane and logical planning which must be done.

Others beside you are being developed to the point of readiness to try to convince the leaders of your world of our reality and our good intentions.

Our stake in all this is exactly the same as yours—*continued existence.*

We are not prepared at this time to reveal how many of your people are standing by, ready to tell of their experiences with us when they are told to come forward.

This is a vast and tremendously vital undertaking. You are expendable. This would simply mean not using you any longer.

You may drop out of your own free will at any time, but this is such an adventure as you will never again have a chance of participating in. You have free will and free choice at all times with no resentment on our part, but much regret if you choose to leave.

July 25, 1978—9:30 P.M.

Now I was told verbally, in my head, that I was to be questioned by the highest body of authority on their planet, which had convened for the sole purpose of putting me through my paces.

We have been waiting. You are under suspicion of.... [I never did know what that was.] You have been disobedient.

"To whom?" I asked.

To us.

And you are who?

The Lords of the Planet. Whose vassal are you?

I am no man's vassal. I work for God.

Which god?

The God who is everything.

You work for everything? Answer.

He who encompasses all.

That one! Encompasses all. All God. We know. It is important to know what you think God is.

It is more important to know that God is.

Excellent response.

I waited awhile and nothing more was said. "What next?" I questioned.

Nothing. That's all. No, wait. I am to ask some questions while we are waiting for response from the Tribunal.... Do you wish to work with us? ...We are. Our purposes are honest and beneficial to all. ...Good. You are specially conditioned for the work. You are *not* indispensable. You are expendable. Do you understand?

You do not yet know your entire purpose for us. It will be told when we are assured of your integrity and good intent. What proof can you furnish of these? ...Your past life does indeed verify them. We must be cautious. You understand? You are right. No one may be permitted to do aught that would put our purpose in jeopardy. We cannot be discovered yet. Soon, but not now.

Can you get information for us?

...Good, we certainly do not want you to commit treason against your planet, your country, or any fellow human.

We must be secretive. It is too bad that it is so, but we must be careful and silent. We cannot know all the reactions of your people to UFOs and contacts made. We need to know how we appear to Earth people. You have the most facility in delivering information to us. What we need are reports of UFO encounters as they are published in periodicals and books. Will you read, garner, and help us to gather this material? ...Yes, as you have already been doing—and it has been excellently done.

We do appreciate your splendid cooperation and your help has been of vast benefit to our cause. Never be nervous for a moment that you are betraying your own kind. This is only to let us know what knowledge Earth people are being given concerning us. Too much has been concealed that would have let us appear as we are—friendly and helpful to your people. They just have not been told....

Will you please wait a few moments. Read or something?...

Don't go to sleep yet. Why do you smile? ...All right, some questions.... The tribunal is back in session.

Are you afraid of lightening?

No. Thunder.

...Good. Are you afraid of spiders?

No.

...Good. Are you afraid of men?

Yes.

...O.K. What do you fear most?

Displeasing them.

What does God want from you?

To love Him and do good.

How do you know?

My heart tells me.

How do you decide on what is "good"?

"Good" is what will not do others harm.

Do you love any man other than your husband? ...I see. As far as you are concerned there are no other men? Is there not someone you like very much?

Could you marry a man from our world? ...Why not?

I pointed out here that I believed any reason for such a marriage would include children, that I personally was too old and also had a hysterectomy, which precluded any resurrection of abilities.

> Oh, God, how stupid! Thank you for your absolute frankness.
> You wish to say something? ...You are supposed to be "mixed up." It is part of our plan.
> Your anger today was uncontrolled. Why?
> ...We recognized the pattern. But you were extremely angry.
> ...We know. But why?

I mentioned here all the malarkey that was going on verbally—day after day being told one thing one moment, only to have it completely refuted and be told something else the next. I supposed it was all testing my various responses, to see the limit or threshold of my endurance in certain ways. Though I believed myself cognizant of the purpose, it drove me nuts just the same, and I thoroughly resented it. I had no control over the situation, though they reassured me endlessly I did. I asked them to leave, begged them to leave, commanded them to leave. Always they agreed and said good-bye, and the next half hour or so were right back.

> Are you still angry?
> ...But you are willing to work with us?
> ...Yes. Sometimes we cannot explain ourselves.
> Do you know who we are?
> ...Yes. The judicial tribunal composed of the Lords of our Planets.
> The rulers of our planets.
> You still dare to tell us, "If it is done my way?" How dare you!

Here I said something about two alternatives: they could either be everything they declared themselves to be, or they could be trying to take over the Earth for their own purposes.

> Do you believe we have such power?
> ...Why?
> Do you know we could annihilate you where you sit?
> Why don't you beg?

"For what?" I said.

> ...Mercy.
> ...Yes. We seek Justice. Very well answered. A little arrogant. Read for a time, our answer is pending.
> Do you believe the tribunal is real?
> ...Very good. An honest answer. Our decision has been made. You are to be dismissed. If you do not believe in us, we cannot work with you.
> Do you wish to speak?
> Is that all? We do not understand your complete lack of reaction.
> ...No, we cannot reveal ourselves yet.

We will write in a few minutes.

You have been more than patient and kind and extremely helpful. We cannot reveal ourselves now.

Will you put your book in order, Parts I and II, and wait for Part III, which will reveal many things?

In the morning a new collaborator will be introduced. He is all business; no nonsense. He will work you hard.

I can't say you passed the tribunal with flying colors. There was a lot of argument. Your answers were too cautious. They found you too skeptical. ...They understand that, but it hurt their pride a little that their veracity was not self-evident.

...I know you have no evidence. ...No. Oh, God, yes. Go ahead on faith, my dear—that is all we can offer.

Sleep then...*Mordalla.*

He continued:

Our lack of understanding of the true condition and intelligence of Earth people has led us astray in trying to make contact with them.

Sometimes it seems the whole endeavor is too costly and should be given up. We have factions here that resist continuation of "Project Earth." We have other factions that feel we are too slow, and still others that feel we are too abrupt. Only very slowly are we working out a compromise solution acceptable to all factions.

Making and maintaining a working relationship with Earth is what it is all about. The first essential has been direct communication. With this writing we are attempting to reach all the everyday people of your area and tell them of our needs, hopes and expectations. Every day we are making new contacts and conversing with them on better terms, due to the success of our communication with you. It has given us new hope that our desire to come into your world as friends can be soon realized.

We know that we must prove ourselves before we may come. We are beings of honor, decency, integrity and loyalty to the cause of humankind—for we, too, are human. There is so much knowledge, particularly in the sciences, that we can give you.

There is so much help you can give us. We may each save the other world from calamity and disaster.

First, we need to establish sane communications, and by this writing we hope we have opened a door.

We have worked with you night and day for more than four months—mainly to develop your abilities of telepathy. So you have learned to detect various personalities, one from another—various tones of meaning and subtleties of expression—and, most of all, to detect a "phoniness." You have become quite adept at differentiating truth from falsehood, which is a factor greatly needed in any future work. You have had a terrible amount of stress and strain constantly all these months, caused by all the various factions invading your mind. You do not know how to shut them off, but presently you will be taught.

We know you need some concrete evidence of our existence and our intent. We know you will not take part in anything of a dubious or evil nature. Evidence will come.

Because we are all able to use verbal telepathy, and you do not know how to shut it off, you have been in contact with the following:

(a) energy existences, as previously described;
(b) existences of soul and spirit, like yourself, who have never lived in your world;
(c) existences of soul and spirit, like yourself, who live on various planes of being, never on material planes or planets;
(d) other humans, greatly advanced and adept psychically, who live in your own world;
(e) humans, like yourself, who are psychic masters from other physical worlds or planets. (These are the originators of the UFOs.)

All of these various personalities or existences, all very adept at telepathy, have been busy with you for more than four months. At first working to cross purposes, or just experimenting, we have all come at last to a concerted effort to develop and utilize your particular abilities for our conjoined effort to stave off coming disasters for your world and some of our worlds. Because all creation is joined, we would all be affected by any catastrophe happening to any one of us. Several factions have been interfering with each other's output, thereby confusing you and the other factions. Oral telepathy was presumably stopped, but no one really stopped. No wonder you were constantly complaining and demanding explanations!

A voice, calling himself Lespora, soon began a most interesting discussion:

Many things are left unexplained. We think it best to clean up all the loose ends that leave wonder, doubt and suspicion in your mind. We have made many errors of judgment all the way through, but, with your gracious patience, we have been able to rectify most of them in time. A few, however, proved disastrous to us, and a pain in the neck to you.

I do like your idioms. They are so graphic. At all times, you have cooperated splendidly and been a good sport. (Once or twice, you did chastise us severely for getting out of line. We deserved it and more, too.)

Now the time has come to clarify some mysteries.

One of the things we have not told you: It was from our planet, after we had devastated it with a blast of power—not atomic, but analogous—that some of our fair people set out for Earth and established themselves on the continent you call Atlantis. There, eventually, they managed to destroy their home once again by fooling around with super powers of a still different kind. We, who were left on the home planet, had no desire to visit a far-off colony called Earth, though we kept in contact with the true Atlanteans as long as any lived. They scattered throughout the globe and eventually became absorbed in other races and all knowledge of their former power and glory was lost.

The.... Oh dear, you won't believe this! The last true Master of the Atlantean race wrote out the greatest secrets in a code and placed them in a secret place. On the occasion of your first experiences with the pendulum, you were given a mental task that was meant to impress your awareness on three levels, and it revealed the place of hiding of the lost records. They—whoever "they" were— told you the secret in the chant that you tried to repeat on all levels of awareness, like a roundelay. You said:

> Off the Island of Crete
> In the Temple of Poseidon
> At the bottom of the sea.

And that is where they lie. Eventually, people from your planet will find them. We have ways of seeing, and they have been recorded and are being decoded.

Of course, neither you nor anyone else can believe this now. But it clears up for you the mystery of the chant, and our elation when you were finally able to tell us about it ten years later!

We, too, had knowledge of most of the secrets given in the code, but some of the Masters departing from here had specialized knowledge with them, and—most importantly—had adapted and learned many additional secrets relative to Earth life and usages. Some are protesting that I should not reveal this—as though your people might think the secrets should be theirs alone, since they were extended or further developed on Earth. But without us, the secrets will have no meaning. We are the ones who must not only decode them, but must teach how to use them. Without us, they are worthless. Therefore, I tell you this over the protests of many of my colleagues.

I almost wish the business of finding the lost records has not come up, for....

That was the last of Lespora. Evidently he was caught delivering material others did not want revealed. He did not appear again.

July 27, 1978—9:36 P.M.

On this day, they gave me some information that I was not supposed to reveal at the time. Now it can be told.

> We are a secret group, dedicated to keeping any country or alien faction from gaining mind control within the United States.
>
> We have worked many years at this, knowing that, sooner or later, the psychic researchers would learn this, or scientists studying brain cells and their activities would learn how to excite them with radio and similar waves. Fortunately, the scientists have started it. Russia is far

in advance of the United States. Had it been psychic researchers or practitioners, it would have been more difficult to combat. Russia is beaming radio waves at many places on Earth. The two the paper reported are the only ones discovered as yet, but there are many more.

Now for a real surprise: *The contingent you are talking to right now is from good old planet Earth!* From where, we cannot say. We have members scattered widely throughout the world. We cannot tell you more about ourselves at present.

So sorry to put you through so much, but it was necessary were you to be of any value to us. Now we know your breaking points. This is a very complex situation, not at all as directly simple as it sounds.

You really gave us some hot and heavy lectures, which amused us enormously. We are really very honorable gentlemen! But it pleases us enormously that you were of such a highly disciplined and moral character.

Ravings of a Skeptic

August 12, 1978—10:45 P.M.
Once again, my friends wished to communicate with Bill:

This is for Bill: For the past several months, since we last wrote to you about Ida receiving messages from certain energy existences, unseen and unknown to your usual world, Ida has been in telepathic contact with several different factions. This telepathy, received by her, has been going on night and day for more than four months. It has been the reason for her tension and appearance of concentration on matters other than those directly in front of her. It has been a terrible strain on her nervous system and has made her absent-minded and seemingly thoughtless at times. Five different factions contended for her attention at first, some of them quite unknown to the others. Then all became synchronized and the strain upon her was considerably lessened. She was extremely confused by contradictory statements. She did not realize that different types of personalities were involved. Since the telepathy all seemed from one source, it seemed to constantly contradict itself, until she could not believe anything and it seemed to her as though someone were playing a very bad joke on her. Gradually, she came to understand that different factions from various places were vying for her attention. Then some very worthwhile conversations took place. During this time, she was given a lot of writing to do. She has it all down in several rather disorganized notebooks.

Now, lest you think this is just Ida herself talking, we are going to reveal some things to make it apparent to you that we are real. We have tried to reach you by dreams, as this is your "thing," but have succeeded only in confusing you, as you try to interpret your dreams by a book that is only correct in a very elemental way. Do not try to interpret them at all. Take the dreams you have been having in absolute straight context. They mean exactly what they say. Ida knows nothing of this. You did not reveal your dreams and we have told her nothing.

She is going to be given some very worthwhile work to do. First, she will write the story of her experiences with this telepathic contact.

It has been given to her to be done in several parts. The final part will be given to her later. She is to get all her notes organized and wait for the rest. When the time is ready, it will come. After that, she has some writing of her own to do. Notes were given to her by another faction many years ago. Unfortunately, her husband at that time made her destroy some magnificent writing that she probably can never recapture. She should, by all means, be given a chance to do her writing when the time becomes right.

You will understand more about us and about this whole episode in Ida's life when she has been able to put her books together.

Up to now, she has not been able to turn off the telepathic messages she has been receiving night and day for four months. We are now going to show her how to keep them from coming into her mind. A little later, she will be shown how to read other minds. We dare not teach any person to do too much, too soon, for some unscrupulous souls have used these abilities for the wrong purposes. We strongly believe Ida will use them only for good. But even so, we can proceed only a step at a time.

If you do not believe this or are not willing that she should do this very necessary writing, we will have her destroy everything and have nothing more to do with this sort of experience. We have constrained her to be secretive about this up to this point. Now we must have your permission to let her proceed. Believe us, this is a *good* program for the world in which you live. There is nothing evil or dangerous to her in doing it, but we do want your permission.

Bill said the choice was mine.

September 18, 1978—10:20 A.M.

Obviously there is someone behind you, or someone working with you. Sometimes you scared us to death with your sudden insights and your actual vision of what we were doing! Now we realize it was something done for our own good. It does sound childish. ...Yes, you called yourself a decoy and sometimes said you felt like a battlefield. Neither of these was exactly correct, but close. You were a teacher, or more exactly a "teaching field," on which, or through which, many things necessary were taught. We recognize this now, but cannot tell more, as we honestly do not know more.

September 24, 1978—1:45 P.M.

You have asked several times—innumerable times, dear Ida—why all this written material sounds like your own idiom, your own manner of expression. For a long time, we studied your idiom and vernacular very studiously—all of us—so that we might sound like you. ...For the simple reason that we dare not sound like ourselves! We cannot yet blow our cover. Also, we studied your use of words, the connotations and special usages, so that we might be sure we were talking about the same thing when we used the same words. Everyone has their own connotations of words. At first, we were not aware that your oral speech and written phrases are quite different. When we discovered this, we had to make hasty corrections.

You asked innumerable times why our telepathic messages in your mind sounded like your own voice, and was not in our individual voices. One reason for this was camouflage. But that was only part of the explanation. Actually, since the mind speech is not physical vibration so much as energy vibration, the actual sound of our voices would not carry too well. Your brain interpretation would deliver them in tones much like your own. However, as you saw when Jamie played the little game of imitating a Scotch burr for you yesterday and today, peculiarities of our individual speech could come across if we wanted them to. We thought foreign, alien or masculine tones in your mind might frighten you. For all these reasons, we tried to keep our delivery as much in your own tones as possible. Now, does that clarify a little? …Good. We'll see you again… *Victor, Amorto, Jamie.*

—2:00 P.M.

So now I come to this moment in time when I am to consider this part nearly finished. There is a great deal not explained that I would like some kind of an explanation for, but since they are not able at this time to tell the truth about things, there is certainly no reason to listen to many evasions which would only have to be clarified later. Therefore, I ask no more for explanations.

So much of what has been given has been later obscured by other statements that do not seem to coincide. But I have the feeling the whole project is of such amplitude that explanations given so briefly cannot be true of the whole scheme of things. They can be only a partial truth—and when another facet is being explored, another partial truth seems to contradict the first. They tell me Satan, Hell and demons do not exist. *Who then threatens us beside ourselves?*

September 26, 1978

I had hoped this part was nearly finished, but today I am told to write my own resumé, reactions and analysis of what has happened, what it all means. I suppose this is mostly to see what I am going to say about it all, to answer questions and that sort of thing.

Well, I will try.

September 28, 1978

Trying to find a pattern of coherence in such a seeming hodge-podge order of events is difficult. Writing, coupled with the constant inner talk, covered about two thousand hours during this second pendulum episode (counting the preliminary pendulum work and discounting hours of sleep). I have sometimes wondered if more goes on while I sleep! Sometimes, when I awoke in the morning or during the night, a kind of commentary—not dream—came to an abrupt halt. Or was it a dreaming? I can very seldom remember my dreams. Another question….

I was asked to sum up what I thought had been going on and how it had affected me. Were I disciplined in any of several pertinent sciences, I would probably have some technique all ready and waiting. Since I have no ready-made discipline, all I can do is write as myself. Frankly, I have reached no absolute conclusion about any of this. My belief/disbelief is in a kind of equalized suspension. At this moment, any statement would be purely emotional.

It has been interesting, exciting, absorbing and sometimes terrifying. Although the events of this second episode were never as wild or as terrorizing as those of the first episode ten years ago, there was still tremendous emotional anguish throughout. Perhaps some good psychologists or parapsychologists could go through some of this and find a coherence and meaning I fail to grasp. "They" never allowed me any kind of certainty on anything. No sooner was something said, than it was revoked, discounted or contradicted. I had no sturdy foothold anywhere; everything was slippery.

I do believe these experiences contained a kind of reality. It was not, as contemporary jargon would say, a "mind-construct" (whatever that is). I believe it was a real event with a real purpose. I don't know whether it was good or bad. I have no absolute convictions.

I wonder: If this episode has cost me two thousand hours of intense and sometimes agonized effort, what has it cost "them," whoever they might be? Certainly more than one personality were involved: Victor, with his sentiment *("Dear Heart, Dearest Girl," etc.)*; Jamie, with his delightful Scottish burr (though sometimes that burr crept in elsewhere and made me wonder); Amorto, with his intelligently expressed concern, his charming little verses *("Under the stars, Under the moon, Ida will shine very soon,"* and *"You will even be happy. I sneaked that in for you...Amorto.")*. Oh, how real Amorto's little verse and later comment seemed! How appreciated—like running through a frightening forest and suddenly meeting your brother, face to face. Would a demon, a thought construct, an archetype, or a secondary personality have worked in those little touches? I can't believe that. Those two touches were what made the whole thing seem real, and bearable!

But if I spent two thousand hours, how many hours did all the others work—planning, preparing, presenting? If they were real at all, they must have spent thousands and thousands of hours! All for fun and games? Not likely!

"They" said they were not controlling my mind—that I always had free will, and that they dared not interfere with that. I did *not* have free will. I begged, I pleaded, I wept, I shouted bad names, I demanded, I commanded, "Let me go!"

"Don't panic. We are friends," they assured me over and over.

Did one of their stupid, extravagant promises ever come true?

How many times were they saying, *"Dear Girl, Dearest, Dear Child,"* and such things, and somehow, underneath, I could hear muttering: *"The glutton is eating again,"* or *"The bitch is not cooperating,"* or *"Oh, my God, is she* never *going to be finished shopping?"* And Victor's favorite name for me underneath the surface endearments was, *"the little turd."* Each time I pointed out these "slips," they were profusely apologized for. Certainly I was not supposed to hear them. Once I heard, "If she only knew what was going on here..." All such slips, and the questions they aroused, made me determined to get away from them. But I could not. "This is *not* mind control," they said, over and over. What was it then? It was certainly *person* control!

"There is a reason for everything," they said. Granted, there had to be a reason for everything. But for whose purpose? Not mine, certainly.

And there definitely were opposing factions! Often I complained that my mind was being used as a battlefield between opposing sides, or several sides. That is why the writing and the daily conversations were so confusing, so inconsistent, so contradictory. On the battlefield of my mind, battles were being fought that I did not understand. I did not know the combatants, I did not understand the issues, I did not have a lot of interest in it and nothing at all to say about it.

Not only my mind, but my emotions were controlled, or at least artificially provoked. It was mind bending—and by adding the emotional context, none of it provided by my own motives, I certainly did *not* have free will. Whatever the purposes of the combatants, they were not only using me against my will and understanding, but were placing upon me responsibilities that I should not have, since I could not control in any manner the events that perpetrated them.

If these are the kinds of minds and characters that are setting themselves up to "save the Earth," I wonder if the Earth might not be better off stumbling to its own conclusion? I do not want to be part of their plans. I repudiate it totally.

However, this seems to be only one very small contingent—one small, stupid, arrogant, high-handed faction (with many factions against it). I fell into the wrong hands.

Another faction realizes this. The following message illustrates this point.

September 29, 1978—7:45 P.M.

These fellows have used you rather badly, and are just waking up to that fact. So intent were they on their own purposes and needs, they were not too concerned about you. You have pointed out your expenditure of about two thousand hours—which is not enough, but we will use that. Of course they were working also—but in relay fashion. And the whole episode was for their purposes and their benefit, actually. Only incidentally, as far as they knew, were you deriving any benefit from all your hard work and distress—and yes, sacrifice.

They are being called to an accounting. Do not be distressed by any seeming purposelessness. There is a reason for everything. *Always!*

As stated before, you were being used by a greater faction to teach them. That has been accomplished.

Now the greater faction has given them instructions to make some kind of retribution or to be removed from "the group" for all time. They have been too high-handed.

...Of course you are confused again. Confusion stops *here*.

It does? I hadn't noticed.

Quite right. We stood aside to let the others reveal themselves through you. Unwittingly, you performed a mighty task, allowing them to spread out their stupidities upon the arena of your mind, where you proceeded to combat them singly and in groups, all the time admonishing and reprimanding and trying to straighten out their muddy thinking. As you did this largely on your own thinking and analysis, you were at the same time developing your own awareness and discipline, which has made you a much more comprehending, diversified and capable person. This is why we allowed this nonsense to continue, perhaps a little too long.

Very quickly you will understand and perceive more.

All of this once more weighs out and balances the negative and the positive. The "teaching" was not easy, but there is no reason why it should have been. I was being taught at the same time. The hardest learned lessons are the most profound and the most valuable.

At times I was inspired to certain speeches but I knew these speeches did not come from me. They were interposed over my thinking, another kind of mind interference—though originating, apparently, from another faction (the teaching faction). It was still a form of *mind bending*.

This constant conversation and writing entered into my daily life to such an extent that I forgot and neglected many things I otherwise would have done. My flowers burned up in the garden for lack of attention and water. The strawberries dried on the vine. I did not listen to people talking, for someone else was busy in my mind with incessant talk-talk. My bookkeeping chores slid six weeks behind. I forgot to call my daughter on Sunday nights—a long-established routine—because my mind was so taken up with "those durn foreigners," as I called them one time. The volume of writing put out in those four months was nothing compared to the volume of chatter. I tried to take notes on some of it, but with the telephone ringing and customers coming in, I was interrupted and before I could get the last down, more was coming. "They" discouraged the note taking. *"We have a permanent record of everything,"* they said. *"You won't need it."* At first I had no thought of a book—until they urged me to write of the first episode with the pendulum of nearly ten years before.

"Why do I have to write that?" I asked. "If you have access to the past record of all my life, surely you know what went on then, probably better than I can recall it now."

"We know the action, but not your thoughts and feelings," they explained. And later, *"We needed to know how the voices you heard ended."*

This left a large question in my mind. They should have known that also, if they knew all the facts of my life and if they were the same ones who had contacted me then.

So many questions went unanswered. One of the biggest was: Why did they ask me many times, both during the first episode and this later one, *"Ida, who are you?"* and even, *"Ida, who in hell are you?"* Didn't they know? They said they could discover all the facts of my life, and still they asked, *"Who are you?"*

Sometimes they said there was someone behind me, someone coaching me, unknown to them. I did not know either.

I find another point most confusing: What part or parts of my mind were in operation? I was not in a trance, as I was fully conscious at all times. I was ready to instantly answer any interruption, telephone, doorbell, someone wanting to talk, a sudden recollection. I would have to say, "Excuse me, I'll be right back," or "My mind got to wandering," or "I got off into thinking what I should fix for dinner."

It did seem that I had, at first, hypnotized myself by watching the pendulum, which opened the door for telepathic reception. One of "them" did say there was a slight dissociation in the beginning, which was soon closed by an actual physical motion. *"You are not, however, using your subconscious, but your superconscious, for that is our area of activity at the present time."*

Was I actually using three or more levels of consciousness, as the ones who delivered the secret of the lost records told me? (They said I could chant the secret on three levels of awareness at the same time.) Confusion indeed!

I have been told that I truly did allow the UFO people to see into our homes and daily life. They supposedly saw the flowers in the garden, certain things in Hawaii they had an interest in (such as the dolphins), and our family gatherings. They followed me for days in my activities, thus learning Earth-person routines and activities. They say this was a great, beneficial thing, for the "UFOers" will now use better methods to contact Earth people and will treat them as equals, rather than specimens to be studied.

This leaves me with a dreadful question: Could not all this familiarity with Earth's daily life also allow them to disguise some of their numbers to behave in a very Earthly fashion, and thus penetrate into our midst undetected? (If they are indeed friendly, this should not matter.)

I was unable to escape carrying out any of these requests to show off our world and our daily life. *I was controlled.*

Did those who were so using me remember the declaration of July 21st by the UFO people?

We want to keep you from blowing yourselves out of the sky. Any great upset in planetary position would simply upset the equalized activity of the universe, including us and our helper planets. We cannot stand idly by and watch this happen. We dare not. We must take a hand. We want to do it in a cooperative way, without compelling or controlling, but if men of planet Earth do not listen and take advice, we may have to take compelling action to save ourselves, and you!

You can see how logical our stand is, and how necessary interference might become.

That is the gist of the whole story.

October 7, 1978—7:30 P.M.

Why, when I was writing about the first pendulum episode, did my unseen correspondents prompt me on one recollection only? It was the only time they called anything to my attention. *"Try to remember the man who came to the door one night,"* they said.

I had to really work on that. I had to concentrate and study hour after hour, day after day. It did not come back readily at all. I could remember no man. What man? When? What did I say? What did he want? What did he say? Over and over I pondered. Then one evening, a picture came into my mind. I saw my room as it had been at that time—my chair, my lamp, my bookshelves. They were all rather vague, and I was not too certain as to the positions of the furniture in the room. Still, it was a recognizable picture. I "saw," rather than felt, myself get up from my chair. I saw the open door, myself standing in it, and the man on the porch. It all came back, as though it were a picture I watched. Then I began to recall details: his appearance, his size, shape, clothing, demeanor. I could not recall his face. Had I seen it distinctly, even then?

Was it only this later recall and interpretation that made him seem mysterious and placed a somewhat esoteric question over the event?

So many questions remain: Why was he so insistent that this was the place he was to meet his son? He insisted. Again and again he said, "But this must be the place. This is the way he described it." Why did my communicants prompt so insistently, "Remember what the man said, what you said." I drew a complete blank at first. I remembered my reactions to him and how hard I tried to tell him something. But what? And what was his response? Why was it so vital that I remember?

I could recall he had drawn a portfolio from his coat somewhere, but what was it he had said?

I could not remember, and it took prompting on the part of my correspondents to even break through to some tentative words. They offered some phrases, but I knew they were wrong. "Let me think," I said, "just give me time." And the words finally came into my head. When I heard them, I knew they were right. It was exact. But had I actually thought them, or had they come by some ulterior fashion? Once I had them written down, however, I was certain beyond doubt that they were correct. A little more mind probing and I remembered my exact words, and I am positive these came by my own recall. The more details I remembered, the more easily more came.

Why was it so vital to remember these exact words? "The message you gave him is the most important event of the first pendulum episode." This "they" told me now.

> Before the year 2000, men are going to have to learn to use a new force coming into their lives, or they will destroy themselves. [Not atomic force—that was years in the past. This was 1968.]

Were these words really a kind of message I was delivering? Were they simply the culmination of my reading and thinking coupled with recollections of some writing I had done years before? And did I really feel compelled to tell him this, or was it only that I was eager to find someone to discuss some pertinent topics with and had found no one. He looked and acted like a well-educated and cultured gentleman. Perhaps I was only reaching out for someone I suspected had kindred interest, before he evaporated into the night.

Where did he disappear to so suddenly? Around the corner of the house in front, no doubt, though I could never remember seeing him go. It was quite dark, but the light shone from the doorway. I just lost sight of him. By that time, I was in a state of slow comprehension—not quite dazed, but slow. Everything seemed somehow more real than real. It could have been my mental state that let me lose sight of him. He just seemed to melt into the darkness.

So many questions remain. Why was he walking this late at night? But then he was from "out of town" and the hotels were not many blocks away. Perhaps he thought it foolish to hire a cab for so short a distance. Why would he not come in and use my phone? Was there no rendezvous with a son? Did he just not want me to hear his conversation? Was he shy about entering a lone woman's house to use the phone?

Could it be a strange, mysterious, guided event? Or was it an accidental meeting of no consequence whatever? How so many times I asked over and over again, "Was there anything in that bit about the lost records of Atlantis?" and every time the answer came back strong and clear, *"It is absolutely true!"* And again, "Was there really a contact with the UFO people?" and again, strong and unhesitatingly, *"Absolutely true!"*

These still are outstanding questions in my mind. Nothing evidential has ever been forthcoming on any of it.

Their final statement was that this was a bona fide telepathic experiment. That is contingent on a very human group trying to make the United States aware of the impending dangers of mind control—not only threatening from outside sources, but perhaps already infiltrating into the United Sates. This human group is supposed to be composed of psychologists, parapsychologists, scientists from other disciplines, people engaged in communication services, business persons, doctors, lawyers—all of whom have been disciplined in certain psychic orientations.

The fact is this: If these researchers can penetrate my mind with telepathic messages and, by their clever use of emotion-laden words, keep me dancing on the string for six months, while I try in vain to break away, it certainly behooves us to study *mind control* before the outsiders are able to do the same thing. Or are *they* the outsiders?

October 10, 1978—1:50 P.M.

You have called our efforts sophomoric, childish and superficial. To all appearances, you are quite correct. And it was all done like that for a purpose, which we thought was good when we did it. Perhaps, on reflection, it was almost stupid. We wanted you to have confidence enough in yourself that you could speak up, defend any encroachments upon your privacy, and talk back when such seemed necessary. You did cuss us out good several times, pointing out corrections in our thinking, giving advice and information, and so on. You were very lacking in self-confidence when we started, but you have become a great deal more aggressive, in a proper way, by this time. Our group has been aided by the very finest psychologists, psychiatrists, parapsychologists, doctors, etc., that your country can offer. Our seeming superficiality and sophomoric escapades were all part of the act.

Now you say—and rightfully—what act? The act was to develop you into a very fine and ready-for-action researcher for our "cause." I hate that word "cause" as much as you do. For our "purposes" might be better, since the purposes are ours and you know nothing about them yet.

At no time has a physical human spoken to you in your mind except myself (Victor), the Nameless One, Jamie and Amorto. No one else has the talent yet to do so, though several are on the point of being able to. Sometimes I speak for others.

I do not know the one who told you that you were "teaching" or acting as a "teaching field" for them, even as you were being developed psychically by them. I honestly do not know who this is.

You have many questions, more than I can answer here. In this, we want to make clear to the readers of your book just what has been truthfully going on and why.

The contact with the UFO people was a true contact. The facts revealed concerning their planets and their intentions are *true*. However, as you pointed out in your little resume, they did tell you that if people of Earth did not take action to avoid destruction of this world, they would do so. You must be sure to quote this exactly as they said it. You must emphasize it—not because they would have you do so, but to make it evident, to the people who read, that all the protestations of friendship cannot and should not block out the fact that their intentions are final. Either we act or they will act. Present that—not as a threat (for their friendliness is very real), but as a possibility not to be ignored. This paragraph is the most important one in your whole book. It is true. It is real. The whole episode with the people of the UFOs is real. I did interpret for them. They know English quite well, but use pictorial telepathy.

I have already given you some reasons why I sound like you when I speak in your mind. The main reason is because I have followed and listened to you for many years. I know exactly how you talk. In fact, we two now talk very much alike—only I am much more profane.

Our purposes are to develop many persons' awareness as defense against any invasion from any source that uses mind control.

The UFO people can indeed teach such awareness and defense, and they have offered to do so in exchange for some of the things we can do for them. This, however, must be authorized by the proper Earth officials, and the time seems a long way off when this might be accomplished.

In the meantime, the human ones of our little contingent are doing what they can to prepare certain ones to be ready for such information and development. You will not be of that crew. You will write on a series of subjects I am not at liberty to disclose now. You have already served our purposes and will not be used in these developmental plans. What we have learned in our exercises and experiments with you has taught us how *not* to deal with others. You have borne the brunt of the experimental stupidities. Others will be more fortunate because of your experiences.

Do I begin to sound like a more intelligent being now? You told me several times that I sounded like an insane nut. I don't think I am, though sometimes I wonder how I can be so stupid and obtuse and make such frantic mistakes. It is all so experimental. We go forward almost blindly, inspired only by necessity and good intentions and purposes.

Ah, yes, one last question to be answered, then I must go. In time, more questions can be answered, but there is not the time now. You ask about such things as the "energy existences", the Flits [see Chapter 1], and so on. I am indeed a psychic master, and I do have contact with many worlds unknown to most persons. The energy existences and all that are quite real, but this explanation cannot be enlarged upon now.

A Reason For Everything

October 27, 1978—10:15 P.M.

We three, Victor, Amorto and Jamie, first presented ourselves to you as energy essences, which of course we are—as well as corporeal and human (and so are you though you do not know how to separate them). There are also energy existences that do not have corporeal form as you and I, but are nonetheless quite as human. From here on, to differentiate between these energy existences and ourselves in our out-of-body or energy-essence form, we shall refer to them as energy *existences* and ourselves out-of-body as energy *essences*. This will save much confusion.

We have also confessed that we are now on Earth working with an Earth contingent, though we ourselves are UFO-connected (however, from a different source than Planet X. We cannot go into this more deeply now).

At first we were not ready to present ourselves as having UFO connections, until we had tested your beliefs and reactions concerning them. Later, we became convinced that you would do a fine job of writing for us—and indeed, you have well started on that project.

As energy essences, in our out-of-body state, we can come and go invisibly amongst you with all the faculties of our human selves, except for color and odor. For some reason, we cannot detect these. We can observe and hear well enough to "sit in" on any conference, business meeting, classroom, or any place we choose, except a few places that are prohibited to us by our advisers. Therefore, we have been able to "educate" ourselves in your best colleges and universities, and know very well what is being taught and experimented with.

We are not permitted to enter into homes or private places uninvited. We cannot invite ourselves to spend the night in your guest room. Some of the UFO "workers" do not have these restrictions, but are sent on definite errands or missions and may and do appear suddenly within the homes of your people. But this is for necessary and benign purposes (which may not seem so at the time). They will not appear in their true purpose and meaning until the time is ready for this revelation. Now back to us....

We educated ourselves in many Earth disciplines and found, to our great surprise, that, while we were vastly superior in technology and science, we were not all that superior in intellect, and even grossly lacking in knowledge of human relationships. "Book learning" in no way prepared us for family living.

After all these years, we are only now coming to fully understand Earth minds and Earth emotions. By following you, dear Ida, in very close, day-by-day detail, we have seen at last what makes an Earth human "tick." We know you are a unique individual—each of you is— and knowing you in no way prepares us for other Earth people (that is, in close detail). But it does prepare us, in generalities, to mix and mingle and associate intimately with your people in a manner never permitted to us before.

To come to Earth and mingle with your people would take great study, discipline and practice on the part of the natives of Planet X. We three are more fortunate to have a closer relationship than they. But, since I am dictating now in their behalf, I shall refer even to we three as part of the total UFO contingent.

Our way of thinking and living is so alien to you that we must spend a great deal of time preparing ourselves to meet and converse on comparable terms with your people. Our intellect (or knowledge) is not vastly superior, but it is very different. It is alien.

We have pointed out that our individual life span is centuries long. When a person has thought and acted in certain ways for 300 years or so, he cannot change his mannerisms overnight.

Certain ones of our people have been trying to communicate and learn to converse knowledgeably with Earth's fine people for a long, long time. But the distances involved, even with actual contacts now and then, make such study a slow and tedious thing. We can study, analyze and computerize an Earth being, and still not be able to talk to him or to really understand his manner of being, his way of thinking and acting. Human beings are not wholly "computerizable" (if I may make up the word). That, in fact, is the great beauty of human nature—its absolute unpredictability. We can estimate probable reactions and actions, but we can never prognosticate one hundred percent.

The complexities of human relationships—the interactions of two or more persons—makes any situation even more complex.

We have studied many years to formulate patterns of behavior relative to approaching Earth persons on an equal one-to-one basis. We feel we have at last gained the insights necessary to begin making some approaches that will not frighten or overly confuse contactees like you.

Our reasons for the confusion we have created are several, but, for the moment, we will name only one: *We were not ready to reveal too much about ourselves.* We wanted to make Earth people realize we were here, at hand, but not let them see too much evidence until we had better plans of approach, verbally and emotionally, worked out.

Now we come to the time that we must advance evidence of our validity. In the stories and articles you read, the question constantly pops up, "If the UFO people are so smart, why don't they say something intelligent?"

It would be of no value to give you scientific principles, laws, equations, formulas you would in no way understand. You could not even write them down correctly. When we have established friendly and cooperative contact with an astronomer, we give him information he can understand and use. It is the same with a physicist, and so on, through the various technologies and sciences. Your own lack of scientific discipline makes it necessary to give you information along quite different lines.

We have decided that the very best we can do for you is to give evidence of some of the reasons for personal contact. Apply these principles to each contact you read about, and you will begin to understand much more. Remember that the published accounts are in no way perfectly accurate. There are several reasons for this:

1. The contactee was too frightened to recall details exactly.
2. Some contactees had hypnotic experiences in addition to their actual physical experience, and do not remember everything or know how to interpret what they do remember.
3. Some contactees are afraid to "tell all."
4. Some contactees have distorted, for their own purposes, events that happened to them.
5. Some contactees are under pressure from associates, peer groups or even social or political groups to tell certain things in certain ways. (It is this reason that has led us to try to contact persons not closely tied to any political, religious, business, or social group— persons who, like you, are independent.)

We have been having conference after conference to determine the best way to progress with the writing. There are so many things we want to tell you, so many books to write with your collaboration (for we have discovered that you have as much to say about certain subjects as we have). That is, we tell it from our angle and you show us how it appears from your viewpoint. This is priceless to us, for no one else has been quite so patient and amiable in helping us discover this new viewpoint, which is unfamiliar to us. We have had contact with humans over the centuries but always from our own viewpoint, always structured by our own desires and purposes.

At this point, Victor interrupted to discuss something totally unrelated:

We are greatly amused at your struggles to understand your fine new camera. May we point out that your habit of reading everything from back to front is really not the best way of understanding instructions of any kind. Please start at the beginning of your photography manual and read it carefully, page by page, getting each instruction under your control before you go on to the next one. Your "hop-jumping" from this page and back to that is only confusing you. There is a lot to learn about your lenses and the various controls. You are taking some fine pictures (with our help), even without knowing what you are doing. But absorbing the instructions of the manual will be a great benefit also! ...And do order the wide angle lens if you want it so much.

Then it was back to work.

Ten years ago, when you first took up the pendulum with such horribly disastrous results, you were first the victim of self-hypnosis and then of an invasion, by us, of communications you did not in the least want. If we had taught you then the methods of avoiding mind control, you would have left off all communication with us and we would not today be dictating this to you. Therefore, we are most unwilling to give you information that would allow you to "escape our clutches" (and here we give a villainous laugh and twirl our mustaches). But seriously, we shall eventually give you this information—not because we are eager to, but because we see that it is absolutely necessary that your people learn this. Already many people in the world are being mindlessly obedient to forces they do not recognize as inimical, or hostile.

Yes, this does seem to be bits and pieces of information. We shall get to a more formalized portrayal of fact later.

We may come in some numbers to effect changes in the social structure of some countries.

You say this is a dangerous announcement, that it will arouse antipathy against us. No, not when it is understood exactly what we mean. We feel that we have an enormous amount of good and benefit to offer the Earth, just as the Earth contains many things to benefit us.

You say I am getting wordy again. Very well, then I shall sign off for a time.

As soon as you have a little more freedom from necessary tasks (getting Bill started in his new shop for one thing), we shall take off with a great deal of explanation for what has gone before, and firm commitment for what is to come after. Take a breather for now, but later today I have a little more to add... *Victor.*

Soon, another took over:

We told you formerly of the conditions on the planet we call Tea Elsta. This is the home planet of the UFOs. Its greatest needs are water and more Earth-type people. There, as frankly and clearly as we can state it, is our purpose in writing this message. We—all the UFO people—do need your help.

On the other hand, there is so much science, medicine, and technology that we can teach you or give to Earth people.

Certain sudden and disastrous changes will take place on Earth before many years have come to pass. Your scientists have warned of these; nothing is unknown. From natural sources, earthquakes will wreak havoc over a large portion of the globe. There will be technological mishaps and sociological disasters.

We are determined to make contact only with persons who can think for themselves, and sometimes we have erred badly on this. Of course, extreme fright can cause an independent person to revert to relying on some authoritative figure or representation (just as you bought a Bible during your first contacts with us as a kind of emotional security). This has happened all too often in the past. It only serves to confuse matters in the minds of the contactees, as well as in the minds of the observers and investigators.

The first pendulum episode, as you refer to it, was a terrible hash for everyone concerned. Ten years have taught us a great deal about how to protect you from other invasion. We had to learn how to handle you without actually doing so. We had to make progress in using your facilities without encroaching upon your rights or invading your privacy. I'm afraid we did both, right at first. But we, too, learn quickly, and I think we have reached a point now where you are as free as possible and at the same time a well-developed instrument for our use. At least you don't complain about it too much. Your grumbling is mostly habit.

We select our contactees carefully. The contact is in no way accidental, though it has frequently been made to appear so. Only after long study do we lay plans to approach or intercept a contactee.

Now we can go ahead with other plans to negotiate contacts of understanding with certain personages we have been wanting to approach for a long time. We can give valuable information in many lines of endeavor. We needed to know how to approach persons so as not to frighten, alarm, disgust, overly excite, or make enemies of them. Bit by bit, we have worked out many tentative approaches for different types of persons.

We still find it necessary to avoid armed or military personnel, for we are fearful that they will shoot first and ask questions later. This has been done too often!

Knowing the Earth languages was simply not enough preparation for making friendly contact. Our life and thought is so alien that we had to learn the beliefs, ideas, traditions and customs of many diverse peoples and nations on Earth. Some of them are so alien to each other as to be unbelievable! When the diverse countries of the earth can neither understand nor tolerate each other, how could we, so very alien, hope to approach in immediate friendship and understanding? The problem of how to approach alien beings without arousing fear or distrust is extremely subtle and complicated. Even now, after all our converse with you, you do not entirely trust us (and rightfully so!). We have thrown you into utter confusion and doubt so many times and misled you and distorted the truth until it is a wonder you even speak to us or allow us to dictate these lines to you.

You say you have come to believe in our good will and purpose and that makes us very thankful, as well as grateful to you, for your continued cooperation and industry.

In our contacts with Earth persons of other nations and other races than your own, we have tried (or my allied workers have tried) to find those among them who would be patient and amenable. It is most difficult to find stable, steady, good-natured persons in any country willing to work so long and so blindly, so capably and industriously. This has held us up to a long, long effort. It has not been quick and easy.

Many persons contacted "over the globe" (as your writers say) have gone galloping off in their own directions, not heeding our pleas for patience, caution and exactitude in delivering our various attempts at communication. Few are willing to lay aside their preconceived notions and allow our lines and meanings to come through in the purity of our intent.

It is heartbreaking for us to see someone we have spent many months, even years, trying to develop into a good communicant go racing off in some oblique fashion, totally ignoring the real message he or she was supposed to deliver. So much sustained effort, minute by minute, working over long months—then it all goes for naught as they take off in some strange purpose of their own because of outside influences or interference. You have been most steady—and, yes, loyal—but many have not.

Many demand from us what we do not have to give, and when they discover we are not miracle workers or the spirits of their dear, departed grandfather, they leave us in scorn. And that is why we were so long in telling you exactly who and what we are. We did not want to lose you as others have been lost due to our telling them too much too soon and too unwisely.

There is a reason for everything. We have said that so often it became our motto! You are not always too patient with us, but certainly we can understand and appreciate your eagerness to get on with it, to get something substantial, verifiable and "real." We are working toward that end as eagerly as you are awaiting it.

Many times we, the communicators, have wanted to tell you all about ourselves, even our real names we use here on Earth (for certainly we are not Victor, Amorto and Jamie. These are names we use for you alone).

So too, contactees who are supposed to be talking to "Ezno from Saturn" are doing no such thing, but talking to one of our fellows from Planet X or Source B, neither of which we shall name more exactly at this time. And believe me, there is a very good reason for that!

Names mean absolutely nothing at this point, so just go on calling us Victor, Amorto, Jamie and Mordalla from Planet X and the Nameless One.

If we told you where Planet X hangs in the sky above you, you would not know how to look for it. Leave all that for the astronomers when a good, agreeable one comes along.

The same applies to our mode of propulsion of the UFOs. This seems to be another question your people frequently ask. If we told you, you would not understand it well enough to write it down. And an error would only throw you further off the truth. It deals with anti-magnetic force. That is the most we can tell now. Again, we wait for the proper person to tell this information to. And what would he or she do with it? It could be used for vast harm to yourselves—and, yes, to us too! We must find the right person and nurture and train him carefully for months, maybe years, to learn how he would use this knowledge. It is not something to be tossed out indiscriminately. And, as you observe, it is not something you want to be personally responsible for. The same factors of greatly needed care must be observed in all the information we can impart. Who is capable of using it and to what purpose will he put it? How carefully have we studied and followed you for how many years? Even you do not know that. And the information we give you has little danger to you or ourselves. We have waited years to impart even this meager information. You have over and over again shown yourself amenable to our needs and desires, but wait patiently for our instructions. (Well, sometimes not so patiently. But your impatience is all verbal; you do nothing to sabotage our work.)

We can no more hurry events than we can hurry your Christmas. All must mature, in proper and orderly fashion. This is such a tremendous undertaking, and involves not only your world and my world, but other worlds as well. So many are involved. So many plans are going forward at once. As you observed once, the whole thing is a geometric system of events, not a simple straight line of progression (as from cause to effect and to effect from effect and so on). Geometric is right! And that will be gone into and explained patiently at a later date.

October 29, 1978—4:00 P.M.

We have observed that our line of progression is not linear, but geometric; therefore, we must not try to piece out any step-by-step diagram of our forward intent. This is impossible. A formula could do it, but would explain nothing to one like yourself who does not recognize mathematical formulas. Therefore, let us say we can best explain our current activities by stating their reasons rather than observing the actual actions—at least at first.

First, we must emphasize that the overall reason for the UFOs coming to Earth is, as we told you, to warn of dangers involved in the rapidly expanding technological abilities of your industrial and scientific promoters. Atomic powers and others which you will soon discover *must* be controlled by a world force if the whole earth is not to be shattered and the equilibrium of the universe upset. The results would be so far-reaching that the mind stumbles trying to phrase it.

The secondary reason for contacting Earth is to warn of factions now practicing mind control, which is already virulent in certain areas on Earth. The days of conjecturing and imagining are over. Mind control is a fact and is being used in subtle and devious ways. No news of this escapes into the newspapers or other news media, nor into magazines. A few books have warned of the dangers, the immediate at-hand danger of that which is already amongst you: mind control.

These two are the most pressing dangers we would warn you about and guide you from, for if they overtake you, you are lost and so are we! To preserve ourselves, we must activate the Earth to protect itself.

We have some reasons more than to give warnings and aid. We need your help, too. Our home planet is without water. We need certain plants and vegetation from your desert regions to see if we cannot grow them in our desert and rocky terrain with a minimum of carried-in water. Eventually, after a great many are established, the moisture they release into the air will condense, clouds will form and rain will fall. Does that sound improbable? It could happen. The desert, the rocks, could come to be covered with living plants in the long, long future. We have time. We can work and wait. But we need the starting vegetation now.

We have another reason for approaching Earth. Our racial members are few. Less than a thousand of the original type of people remain on Planet X. (This is not true of the Source B.) Planet X needs new life stock.

Our heritage is from the Earth. This was hinted to you long ago and then the subject was dropped. Our people came from Earth many thousands of years before Atlantis. There were great civilizations on Earth. Did you think the use of electricity, television and radio were new? They have been invented (discovered) over and over again. Airships and space travel are *nothing new.* Humans have gone from Earth to the moon and far beyond before. We came from the Earth multitudes of centuries ago. *We* were Earth's pioneers in space.

Then, after centuries, we managed to devastate our planet, as we described. Some went back to Earth to join Atlantis. Some stayed hopefully on. History repeats itself into boredom.

There are those factions among us who want to give up here and return to Earth to live. We are so few. We could mingle among you and cause no ripple in your society. We would simply "fit in."

Others do not want to give up the kind of life we have made for ourselves here.

Factions! Factions! Factions! And there are still others with more ideas! And this is the reason our approach to Earth has been so indecipherable to you. We are seeking a dozen different things.

Some of the navigators, who are out of Planet X stock, seek only excitement and adventure. They deal roughly with your pilots and planes. They tease, torment and cause havoc. They are not good advertisement for the serious causes of the rest of us.

If we were all one contingent, all working for precisely the same purpose at the same time, if we were only synchronized and cooperatively organized, we could have approached Earth and made friendly overtures many years ago. We cannot "get it all together."

We, the communicators, Amorto, Jamie and Victor, whom you have known under many guises, are under instructions from what is probably the leading contingent. We have been trying frantically to get cooperation for the other factions to all work at the same time for the same ends so we could proceed in an orderly fashion. This is the problem with everyone being "so intelligent." Everyone wants it done his or her own way. Too intelligent to cooperate! One begins to wonder where intelligence ends and damn foolishness begins.

I, Victor, am supposedly coordinator betwixt heaven and Earth—that is, between our planet and planet Earth. I am trying to get Earth people to contact their people and get a good thing going to help us and help you. At this point, I don't know who needs the other most.

At all costs, we who can must try to invoke others to think constructively about the great needs that lie between us.

All the hocus-pocus some of the contactees have told in the past is just revolting to the sane and sensible people of the Earth. And to us also. We must counteract this foolish barrage of garbage with some straight-out facts about ourselves.

I feel this so fervently that I am making your hand write faster than it has ever done in your life. I hope you can read it to copy. Dear girl! We need many more like you, so patient and willing to do a good turn for a neighbor, no matter how distant.

Of course, we need to contact your people on a more sane and solid basis. All the camouflage given up to now has only served to alienate those we wished most to reach—the thinking people.

Our reasons for this camouflage were several, but principally, we had to hide ourselves and our intent until we could thoroughly study the contactee. Would he use our information for his own ulterior purpose, or would he work for the good of all? What were his capabilities and limitations? And so on. So many questions to ask and be answered—and so few would give us the time and help to find the answers.

We have been exploring many methods of learning to converse with people, and find there is only one way—the way of truth and straightforwardness. On some things we just have to trust to luck—or more exactly, trust to the probabilities of the correctness of our analysis of the individual. Hopefully, we have analyzed his or her probable actions and reactions somewhat correctly. Here the unpredictability of human behavior is the troubling factor. We think we have analyzed someone to the last probability and he up and betrays our confidence in one manner or another. One other person can lead a contactee so far astray from our avowed pact that it were better it had never been made. That is why we impose caution and discretion on a contactee. He, himself, may be thoroughly convinced of our integrity and purpose, but how can he convince his father, brother, employer or anyone else of the validity of his experience, or the veracity of our statements?

Or perhaps a woman reads an article that arouses fear or doubt and throws her off the whole project. She may forget that the article or book is sheer conjecture, while our presentation was fact and reality. Not only is the article conjecture, but too often it is written for the sole purpose of selling a story, and the facts are distorted in the name of reader interest. Or again, it is written by some know-it-all or smart aleck (as you would say). Or a well-thought-out article is manipulated by a too-busy editor into something that was not intended. Yet all of these stories prey upon the credulity of the contactee until she knows not whether she heard us correctly or at all, or if she should have listened.

As you see, our problems are manifold.

People ask for evidence, something concrete to prove who we are, that we are what we claim to be, and that we intend what we say we intend. What kind of evidence would convince them? Thousands of sightings of UFOs have not convinced the majority of the population of their reality. Pictures and carefully investigated stories of contactees have not been evidence enough. What more can we give? We want to know—need to know—what we can give and what we can do to prove ourselves.

We have tried to contact persons of great rank and high authority and the stories of our contacts have been buried. Out of caution perhaps, but buried.

We have tried to approach army, navy and air force personnel. We were shot at without question, chased by planes, forced into maneuvers that endangered both the plane and ourselves, and then blamed for the action.

What can we do to prove that we are who we say we are, that UFOs are from a planet in a far galaxy and another source too, and that we intend what we say we intend—to aid Earth in the proper use of atomic and other more powerful forces and to teach how to avoid mind control, which is already in use. What can we do? What evidence can we give? Please help us!... *Victor-Mordalla-Hweig, Amorto, Jamie.*

November 17, 1978—6:40 P.M.

Tonight we are going to let you in on some things that we do not want you to reveal as yet. But for your own peace of mind, you should be aware of several different facts.

There are many types of entities involved in this project. We have listed some of these heretofore without elaborating on their whereabouts or their conditions of being. We want to tell you more about them, but not at this time.

We cannot give our true names or locations, but we do confess that you have not been given any true name or true location. We have had to protect ourselves in many ways, and for a good dozen different reasons. Your patience is greatly appreciated. Please bear with us just a little longer on the revelation of more details about ourselves, our

places of being, our true names and all of those things that you wonder about.

Our circumstances are such that we cannot reveal very much to
the general populace, but we do feel that you have earned some information that will put your mind at ease about some of the facts concerning this project and about the various personalities with whom you
have had to deal. We realize there is great confusion in your mind as
to who is who. We confess that you were misled a great many times
(and confused still more on purpose, as we did not want you coming
up with the truth too early). You sometimes guessed much too closely
for our comfort. We had to keep you confused. That is the reason for
all the fibs that were told to you constantly. We realize, however, that
you are becoming exasperated with this constant confusion, and we
certainly do not blame you. Indeed, your patience is quite remarkable.

We three, Victor, Amorto, and Jamie, took turns reading a prepared script, for the most part. We used each other's names rather indiscriminately. Whoever was handiest to the script read it.

I, Victor, have interpolated and given some pages on my own, and
will give still more in the future when one large stumbling block has
been removed.

For the most part *all scripts were prepared and given to us by our
instructors.* We, ourselves, did not prepare, write or necessarily endorse what was being written. In other words, we are simply the ones
who delivered the message. We did *not* concoct it. We did not write it.
We were given various personalities to assume, according to the
needs of the moment. Sometimes we even impersonate each other.
…Yes, the Jamie burr was delivered by myself, Victor. It was a little
fun thing thrown in, partly for good humor and partly to show you that
we did not have to speak *only* in your own idiom and manner. To indicate to you that we were indeed separate personalities and no part of
yourself.

The Nameless One does his own writing and is learning to transmit. He is not of our crew, but is a faction all to himself. He collaborates
with us and also works on his own.

Roger has been much too troublesome. It is unfortunate that his
services were necessary for a time. He is not a corporeal human, but
an energy existence, as much a soul as you and I, but never born into
this world. Unfortunately I, Victor, have aroused his antipathy and he
has tried to displace me in your regard.

We believe with a little thought you can now unscramble most of
your puzzlement about various personalities.

Your second biggest worry has been *who* has been with you
since birth, as several have claimed this distinction. We believe it to be
the Nameless One, as he does seem to be definitely acting in the capacity of a Guardian for you. From what has been said previously, we
also believe there are others with whom we have not had much contact to date. We hope they will present themselves and we can all
work in unison—otherwise, we are always afraid they will take you

away from us. The Nameless One has ordered us to enlighten you as much as possible and to tell only true facts from now on. He knows you are going to bug out if we don't come through with facts and don't keep our promises.

...Yes, we three, Victor, Amorto and Jamie, are indeed those who came to Earth in 1940, when you saw the UFO in the desert in Southern California. There is much more to be given on this encounter, but not until you have learned a great deal more of other things.

We have told you honestly how you were checked on from time to time after that, how we tried to impel you to use the pendulum and how we tried in other ways to telepathically communicate to you. Now one more confession or revelation that you have been wanting clarified: We three and the Nameless One are all corporeal human beings, just like yourself. Even we do not know where the Nameless One comes from, but we do believe he is of the Instructor class. The things we told you in writing about Planet X are all true. But one fact we have not told you: Those who give us our instructions are unseen and unknown to us. We know absolutely nothing about them. The Council is on Planet X. The Tribunal is of the Planetary Rulers or Lords and is connected with the Instructors in some fashion. The Nameless One also has some connection with the Instructors, and, though we cannot prove this, we believe it is possible he who is the Head Instructor. He does give us instructions in uncertain terms, and we obey him, for we feel he is some kind of higher order.

Why do we keep asking about who you are? You think we should know that, since we know every fact of your life on Earth, but I must tell you that we feel there is some mystery about you. We have no idea what it can be. Some of the things you do and say and some of your past writing makes us feel that you are in some way more than you appear to be—or at least that some powerful personage works through you. More powerful than your apparent Guardian, who is a very high placed person not of Earth or Planet X. There are so many mysteries we cannot solve for ourselves—and indeed, you are one of them. Who gave you the secrets of the lost records of Atlantis and why? So many times you scared us half to death with your statements or your seeing what we were doing (not exactly, but close enough that it made us hold our breaths).

So, dear girl, you are as much a mystery to us as we are to you. You do not know all the details of our individual lives as we do yours, but somehow you know more about us than we have ourselves revealed. It is more than putting two and two together. ...No, we do not want to upset you or make you worry about yourself, so we shall say no more now. But if you find any clue to what goes on with yourself, please cue us in. We are as eager and curious to find this out as you are to find out about UFOs and us.

The Nameless One has some work for you to do, and will be back in three or four days. He has been gone long and we wish he were here to help us unravel some problems.

Later that same night, a mighty voice woke me:

> *We are of the tribunal*...Lords of the Confederation of Planets. You are not understanding our meaning.
>
> When we said, "You are to be dismissed; we cannot use you," we did not mean you were not to work for us under these others. We meant we could not use you *directly,* as we use them.
>
> The reason for this is no fault of yours.
>
> Your adherence to God and right principles are heartily commended.
>
> We cannot use you because of certain factors in your background that have already developed you and planned for you in certain ways.
>
> We would be intruding on other well-laid and long-developed plans. We were not aware of this before your first meeting, but learned of them only during it.
>
> We regret that we did not make ourselves more clear and that you had the misapprehension that you had failed us in some way.
>
> Your adherence to what you call the good, the true, and the beautiful is very highly commended.
>
> We salute you...*The Tribunal.*

I then turned to my own writing.

My husband, Bill, has read most of these pages. He has a tendency to go to sleep when he reads. He is the world's staunchest skeptic. Mostly he believes, I think, that I am fantasizing, or writing from some kind of inspirational make-believe. If he had gone through, or knew fully, the physical and verbal accompaniments to the written parts of the situation, he would know full well there are countless aspects I could not possibly make up or fantasize. The ready-to-accept part of him reiterates:

"Why don't the UFO people tell you something practical, something we can use? Why don't they give you a cure for cancer, or something?"

To his way of thinking, that would be evidential.

Truly so, perhaps, and, believe me, I have wondered about this also. In fact, I have begged, pleaded, cajoled, and outright demanded something equally evidential. It seems they do not have "release from their instructors" to give out with such.

I can see reasons why it would not be wise for them to do so:

> 1. They say they have no real military might. They do have technological knowledge and psychic powers that can be used in defense. Knowledge is their *only power* and means of defense. Why would they want to dissipate this haphazardly? They most wisely save it until proper contacts have been made and they know our military weapons are not going to be used against them.

2. Again, according to them, they do not have com-
mercial enterprises or use money as a medium of
exchange or garner for themselves wealth or possessions. The only wealth they treasure is, again,
knowledge. The knowledge is their only *means of
barter*. Would we hand over to them ten tons of
gold just to prove ourselves friendly? What are we
asking of them, gratis? Should they be more altru-
istic than we? Do they owe us something? They
need some things that we have. And who is to receive their largess, if given? Someone who will set
up a medical laboratory and make millions for
himself with his given knowledge? Or would they
wait to give their knowledge until they are certain
that all humans on earth would share in it *equally*?

Let us use common sense before we expect them to give us their
knowledge, which is their *only* wealth, and their *only* means of barter.

Were they to give us their only means of barter, they would be left
with nothing but aggression by which to interact with us, should we
prove recalcitrant or unfriendly. By all means, let us allow them to use
their negotiable assets in a good and friendly way—which is the way
they hope to use this, at the proper time and in the proper way, so that
all the people of the Earth may benefit.

I am not "taking their part." I am only trying to understand and to
report what has happened (and what I have concluded concerning this
and where I still cannot come to any conclusions. I am trying to see
from their viewpoint, which is the only way to understand anyone.

It seems to me that other personages—other beings—are involved
in this lengthy episode. (It has been a year since I first took up the pendulum again for a sustained experience.) I have not had enough contact with these other personages to determine who they are really
supposed to be. I do not even know who the Nameless One is. Other
than the perennials, Victor, Jamie and Amorto, he is the one most often
in contact with me, but he remains nameless.

March 9, 1979—9:55 A.M.

So much has been going on—on both sides—since we last
wrote. You have been quite busy with Bill's new shop and with having
your daughter for a visit. Now we must proceed. Time is essential.

First, continue typing on the first parts of this, which will become a
book. Corrections will be made as you type—interpolations to clarify
some of the confusion. The more you can type along, day after day,
without too much lag between, the more sense it will make to you.

Second, a new type of work will soon appear. It will involve correspondence. Do not hesitate to undertake it. It is necessary.

Third, future writing will be forthcoming without all the personal
stuff to interfere. This background work is finished.

It has been most difficult for us and you to come to terms with each other. You do have someone unknown to us protecting you. ...No, beyond the Nameless One, who is a kind of guardian. We have tried in many ways to make this or these unknowns reveal themselves. We have no idea who it can be. ...Yes, that is the reason for some of the problems and travail you have undergone. We do use you as "bait" and "decoy" rather frequently. It is part of your task as instructor.

Your big question now is how to consider Victor, since he is presented one day as a hero and the next as a villain. He is a good fellow and has done a splendid job. I don't think anyone else could "handle" you quite as well! Victor has many talents. He is an excellent psychologist. He is not a nut at all. So you may henceforth trust and depend on him—but personal conversation will be very sparse.

Your next big question is, "Are these four fellows human and on Earth?" Yes, we are all human (as previously told), and at present we are all on Earth. None of us is entirely born of the Earth, but we pass here now as Earth people. No more on this today.

One last thought: You should have some Earth contact or Earth friend "on the surface" to confide in. Your daughter would do, but you do not want to worry her until you know more of what this is all about. Therefore, you will soon contact someone who will serve this purpose. ...Yes, all we can advance tonight is chit-chat, so do your reading and do not despair!...*Us.*

Human Like You

March 22, 1979—4:00 P.M.
A message delivered to me this fine spring afternoon, from the *Instigators,* or "those who make plans":

> We have suffered enough agony over your rejection of us. Please be kind! It is indeed a mutual agreement.
>
> We are disclosing at once a "surface" contact, as we see it is advisable for you to have someone at hand to confer with. There is too much undercover and not enough out in the open; we do understand that. He will be cognizant of similar situations and knows a great deal about who, what, and why. ...Yes, we understand why you underline "at once." It shall happen.
>
> Let us be friends again, and we shall do our very best not to harass or torment you with indecisions and contradictions. You do try so hard to put it all together and we repeatedly confuse you. That is not kind, but this project cannot afford over-kindness. You have indeed earned the right for a more open relationship, and it shall be forthcoming at once.
>
> We not only "forgive" you for your highly spontaneous declaration of this afternoon, but were vastly amused (to the astonishment of those who think us rather "stuffed shirts" and very formidable). We must maintain our position for the safety of all concerned, so please refrain from further demonstrations of this sort. Your point has been adequately made. [I had told them what they could do with their information, etc., if they did not start keeping one or two promises. I would much rather they made no promises at all than to have them forget or renege on them.]
>
> Please accept Victor as your mentor and aid. He would be so distressed otherwise. He is the scapegoat, much more than he deserves.
>
> Continue with your copy work. Your surface contact will be made very soon.
>
> Thank you for reconsidering. We do respect your stand...*The "Instigators" (as you call them).*

The night the man had come to my door, expecting to find his son, "Raymond Montgomery," had always been a mystery to me. One afternoon, I questioned Victor about it. This is what he said:

> I suppose, dear heart, now is as good a time as any to partially reveal some of that mystery. More will be revealed later.
>
> You are right to suspect that the "son" was nonexistent, an excuse. The man at the door—the man in black as we labeled him—was our Jamie. This was the third time in your life you had met him! ...Ah, let me keep that a secret a little longer. When the time is right, before long, it will be revealed.... His coming was for the precise purpose of having you deliver your message, as you felt so impelled to do.
>
> The message you delivered was entirely your own, concerning the mysterious force that would come into the world before the year 2000—a force humans would have to learn to use or they would destroy themselves. It was kind of a "graduation requirement"—something akin to a term thesis. It was to illustrate to some "outsiders" concerned (not the communicators) that you had studied well, analyzed intelligently and reached a somewhat profound conclusion all "on your own." (Truthfully—as we once briefly mentioned—all great ideas come about through collaboration of the person who has the idea with "other forces".)

At this point, Roger interrupts:

> Now for my own true confession: I, Roger, am an energy existence only. I live in a world coexistent to your own, but am highly qualified to aid in your psychic development.
>
> I (in my actual being) and the Communicators (in their energy-essence being) are parallel to the stuff mind is made of. We are energy patterns, which, when active, become fields on which thought patterns, concepts, and analysis take place. We are then almost, but not quite, the same as mind congruent on your mind. We are still individual personalities all the same. It is so hard to get this across. We are *not* "other part of your mind." We are ourselves. But we move in conjunction with your mind when we are communicating. As energy essences or existences, we are personalities—individual, with developed qualities and attributes—just as you are.
>
> Dear girl, this is almost impossible! But we are speaking true. We are in no way part of you. We are not "in your mind." We are more like congruent or superimposed on it, but not exactly that either. We, as ourselves, parallel your mind. We could indeed, when active, call ourselves electro-forces that play upon or communicate with the magnetic field that is your mind. If you had more scientific background, we could make ourselves understood better. (Not faulting you, dear Ida, just pointing out a fact. You were carefully preserved from entering any scientific discipline just so you could keep an open mind. It had to be so. ...Yes, we know full well how much you longed to be an arche-

ologist. Maybe you will still get your wish—but that is a confidence I must not reveal.)

We have learned to bridge the gap between our field of force and your mind through telepathy. We are devising easier ways of doing this with others.

Your experiences must be published, so that others will not be so perturbed or frightened at our coming, but instead, make themselves ready to receive us.

The technique of confusion is perpetrated by the UFO boys, for whom I, Roger, am but the mouthpiece. The communicators only relay what they are told. Although we bear the brunt of your discontent with the "confusion technique," believe us, we do not originate and certainly do not endorse it.

I know you wonder about my jokes about being human. Well, I do consider myself just as human as you! But I am not, and can never be, as *corporeal* as you!

…Ah yes, the joke about me wearing glasses. Although an energy existence, I am a true personality and an individual. I am a psychologist—or perhaps more comparable to a psychiatrist. I do not wear glasses as you do, but I do use a kind of scanner to view the activities of certain beings—*not* in your world. The glasses bit began as a joke and we got carried away with it. I presented myself to your mind (not your brain) as I was scanning something, and your mind interpreted me as a man with wide cheekbones wearing glasses.

This was a physical representation—yes, a thought construct—concocted by your mind as the only way it could understand our energy-patterned presentation. Are you following this? Yes, I think you are.

Your mind interprets our signal only according to *what you already know!*

This is very important! It accounts for the reason of your constant complaint that we only tell you something parallel to what you have already seen, heard, read, been told, thought or believed. We try to get patterns or pictures of a kind over to you, but your mind can only interpret them in symbols it already knows. This is a very big stumbling block to our presentation. It is the *major* reason we asked you to read all you could find written on UFOs. We wanted you to have more symbols of understanding for us to work with. We know you are not altogether satisfied with the reason we gave you for reading so much—so that we could know what Earth people were being told about UFOs. You thought that a very flimsy reason and thought of many other possible ones, none of them true. Here I tell you the real or main reason—to fill your nice little head with much material and many symbols for us to draw on in our work.

We have now learned your word symbols in "Americanese," Western style. That is why you can now receive and write so fast for the most part. Sometimes we still have to present picture symbols for your mind to work on. Actually, this is mostly practice for you to learn to do it quickly and competently. Descriptions do not help us relay pic-

tures of ourselves to you because our symbols are alien to you. Only
if they pertain to something you are familiar with will they be meaning-
ful to you.

You thought you saw a large dog with Amorto and Jamie. We
were sending a symbolic picture of another type of guardian companion entirely. But the closest translation your mind could find was a
large police or guard dog—for that is what a guardian companion of
non-human nature would most probably be. Am I getting this across?

Yes. I understand now why, in the early days, someone told me I
was helping with the translation more than I knew. I could not understand what this meant before.

Good! I am explaining better than I thought!

...Yes, we could arbitrarily plant a picture in your mind, but that
would be tampering with your mind directly—and that we are not permitted to do. The actual picture would startle, even frighten you (and
you have enough qualms about us without adding more). Also, that is
not what we want to do. We want to develop the receptivity of your
mind until you yourself can pierce the shell that separates our worlds.
I speak now of my own world, the co-existent one in which all psychic
activity takes place, not the UFO one. We want to break through the
barriers between your mechanistic world of time and space, and our
coexistent world of psychic being and psychic energies.

We do not know if this is possible to your conscious mind or not.
Perhaps it is not even desirable. We certainly don't want it to happen
yet.

March 28, 1979—12:25 P.M.

Dear girl, we can see, by going over your typed story, how confusing it was to you! We did not take time to differentiate the origin of the
messages received. Even though relayed by myself, Victor, Amorto
or Jamie, not all of the written material originated with us. Since we
were all working more or less together to get the telepathic messages
across, we failed to realize that you could not possibly understand our
differences. Geometric personalities was only one of the confusions.
Actual separate personalities interposing messages only added confusion...*Roger.*

—6:58 P.M.

In our entire organization, we have energy existences, UFO people, Earth human existences and other types of beings of which you
have no knowledge. Altogether we are now trying to get over some
important information that Earth people will someday find useful. The
Communicators contact you through their energy essence only.
Through them you have relayed and been relayed messages from
many types of beings.

March 31, 1979—9:00 A.M.

We do, indeed, have contacts on Earth who are working for our most benign purposes—Earth people, that is. ...No questions now. They are doing research in various lines for us—not in human relationships and human psychological reactions, which is your field. We do not double on any field, for we haven't time. You alone have done this one thing. Others are researching historical events, natural catastrophes, disasters, and so on. You will be meeting some of these others soon.

Now the next revelation—and this is the tricky one: You have been in contact with some mighty spiritual forces as well as with us. We were not aware of your place in their concern until we were well into this work. We knew someone was, as you say, sustaining you—helping you in tight spots and so on—but we did not know who they were. We now know they are indeed from the spiritual area, and can do many things we cannot approximate. They have worked with you since your birth, and have spoken to you several times (though you had no way of separating them from others who have spoken). So it behooves us to mind our manners a little, to be more open and above board with you, and to stop confusing you so mercilessly.

You do have a greater task, not yet revealed, and your work with us is only a beginning of great work to come. They allow it as a kind of training for us both. You have been teaching us many points of ethics and correct treatment of Earth humans without quite realizing it. I fear we have been quite ruthless sometimes in getting our point across. We have learned to be more reserved and gentle. You have been learning some useful lessons yourself—to stand up and be counted, for one.

The "Inspirators," who give internal aid to you and others, will not be with you for a while. They tried to come forward, but there is more valuable work to be done first. In a few years, we may relax our efforts enough so that they can come back and give you more aesthetic work to do. Right now, the practical is the necessity.

Now the third and last revelation: We *do* have some of our members on Earth, but they have not had the intimate contact with your people that would be necessary to learn the things you have shown us about human relationships, family life, etc. They are many and their work is secretive. We cannot reveal more.

...All right, I am being pushed to reveal the absolute truth, so I must. Victor, Amorto and Jamie are indeed in residence on your Earth. It pains me to tell this; I do not think it wise. After all our declarations that it was not so, now to have to confess that it is. Victor tried to tell you, again and again in many different ways, but he was always forced to retract. They are, indeed, all there and have, at some times, been not too far from you. This will never be retracted again; it is the truth.

It was truly Jamie who came to your door that night in 1968, ostensibly to find his son (who does not exist). It was so necessary for you to remember that episode—and we prodded and prodded until you did. He was present only in his semi-corporeal state. That is why he would not come forward into light nor enter your house to use the phone. You would have detected that something was unusual.

So you have seen a UFO person! This was not the only time. You do not now remember, but in time you will.

We are truly sorry to have given you so many false leads and so much extra work. The story Amorto gave you about the UFO on the desert in Southern California is all true. All of you being together in the same area that night was no coincidence. (Indeed there are no coincidences. No explanation on that now.) Victor has been telling you the truth, then having to retract it until he nearly lost his mind. He wanted to be nothing but honest with you.

Perhaps that is enough true confession for this morning. Shall we go to the typing now? Thank you, dear...*One.*

April 9, 1979

I was reading *The Andreasson Affair*[1] by Raymond Fowler, which I had just purchased. Betty Andreasson, while under hypnosis to investigate a possible ET abduction, delivered a message that kept emphasizing "three and four, three and four." Roger had been bugging me as I read, and I asked him if he knew what the strange language words were that she was saying. He said he did, but he thought he should not tell me. So when I got to the "three and four," I cried out (interiorly of course), "Oh, I know what that means!"

You do?" he said. *"What then?"*

I told him, and added, "I wrote about it in my outline so many years ago."

"Do you have the outline somewhere? Could you read it to me?"

All at once I felt very strange, as though I were holding my breath and listening, but I heard nothing. I said, "I'm not sure if I'm supposed to. Let me ask." I didn't really know what I meant myself, but I made my mind very still and into the stillness I sent the question, "Should I read this to him now?"

At first there was no answer, but then a voice said, *"Yes, it is vital he see it now."* I wasn't satisfied. The thought came into my mind that Roger could be saying that himself. There was no way I could tell. So I stayed very quiet, holding my mind in this vast stillness, and waited.

Some kind of unspoken command said, *"Turn out the light."*

I turned out the light over my bed, where I had lain reading, but left the dim desk light on. Then I found myself going through some of the same motions I had, on occasion, used during the first pendulum episode. First, I raised both arms over my head, just a little way with palms together, them lowered them below my chin, then back to touch my lips, all the while with palms still together. Then I separated the hands and crossed my arms on my chest, each hand resting on the op-

1. Fowler, Raymond E. *The Andreasson Affair.* Newberg, OR: Wild Flower Press, 1994 Reprint.

posite shoulder. Both arms went down at my side, palms upwards. Both of my arms crossed again over my chest, hands on shoulders. I stayed this way for a few moments. During this "ritual," I could not make another motion or think; it was a completely controlled and impelled kind of action.

Then the command came telepathically, in its "normal" manner, *"Get your pen. Write."*

This is what I wrote:

> No one knows these secrets who are not of a special order. We joked before. This is real. This is the answer to our questions about you and the question, *"Who are you?"*

Then the handwriting changed. I wrote:

> Believe in this. You are truly so. Be discreet. We are watching over you. You came to tell the world many things with the help of the Nameless One. He will write now.

Once again, my handwriting changed a bit as *One* began:

> Yes, I am the Nameless One—the one you called "The Powerful One." Someone has been using my name to mislead you.
> We came from the same source, you and I, to bring these two worlds together—not our world, but the world the UFOs come from and your present world, Earth. We had a choice, you and I, which of us was to come to Earth. You gave me first choice, as you always do, and took what was left. Sometimes you feel a faint rebellion, but you do not remember what causes it. There is nothing to fear. Be not disturbed by any of this. Much later you will come to know all. For now, sleep and rest...*The Original and only true Nameless One.*

I had been complaining for some time that I did not feel the sense of power that I had at first when the Nameless One wrote or spoke to me. There were no more thuds and bumps in the atmosphere around me, and I had the uneasy feeling something was not right. There were so many verbal confusions going on all the time I could not sort this one out.

PART 3

Expanding Reality

Seeking Out Dr. Sprinkle

April 11, 1979—12:30 P.M.

We are so pleased with this development. By all means, carry through on your ideas. We are so pleased! I can't say that strongly enough. Surely you are guided and sustained by forces beyond *our* knowledge, and we think we are pretty Know-It-All's. To find exactly the right person at exactly the right time! By all means have lunch, then write your letter. Let Bill see it before mailing, if you wish. Consult him as much as possible from now on, so that no dissension may arise between you. We cannot have any jealousies or other troubles of such a nature for you to contend with. Your work will be quite stressful as it is.

Good luck, old girl. We are all with you and for you...*Us.*

The "development" referred to was my discovery of a magazine article describing Dr. R. Leo Sprinkle's work with other persons who were "hearing alien voices." I immediately wrote to him of my own experiences.

April 11, 1979
Dr. Leo Sprinkle
Associate Professor of Psychology
University of Wyoming
Laramie, Wyoming

Dear Sir:
In the June issue of Beyond Reality, *in the article called "The Unearthly Voices in my Ears," I discovered you have been researching those persons who claim to have mental contact with UFO people. I am sure you will receive letters from many others with the same declaration.*

I have had two episodes of mental contact. One, ten years ago, ended disastrously for me, and I guess for them also. Another is an ongoing experience in which I am in daily, constant contact, via mental telepathy, with several contingents, one of whom claims to be of the UFO people. This contact has contin-

*ued since November of 1977. In May of 1978, telepathic writing
began (not automatic writing). At this point, I have perhaps
100,000 to 125,000 written words purportedly from them and
allied sources. There is also daily verbal telepathy. I have prepared about 90,000 words exactly as received with my own explanations as to what was happening and my reactions
thereto, just as it occurred.*

*Those who seemingly dictate this writing tell me to have it
published by popular press, but at the same time to have some
scientific researchers examine either the writing or myself, or
both. In other words, they want the writings to be presented to
the general populace exactly as delivered by them, but in no
way do they object to a complete scientific scanning.*

*If this meets with your research requirements in any degree,
I would be happy to discuss it further with you. Please inform
me how this could be best applied to your interests—if at all!*

Sincerely yours,
Ida M. Kannenberg
(Mrs. William P. Kannenberg)
Age 64
Married
Antique shop keeper 12 years

Meanwhile, my telepathic communications continued:

April 13, 1979—10:25 P.M.
The Hidden One speaks out:

> I do like the new name you have given me since my original one,
> "The Nameless One," was usurped by another. As you guessed or
> detected, he and Victor are quite inimical. This is one of the reasons
> for the constant discrepancy in their account of Victor, his personality,
> his reality and his proper place in your training and development. ...By
> all means, accept Victor's account as more accurate. A few exagger-
> ations and a little teasing are his only misdemeanors.
>
> The other party has been more careless with the truth and has
> caused you much worry, anxiety and distress. He is no longer with us.
> Question not his identity, personality, other names or what has be-
> come of him. He is a nonentity as far as our future work is concerned.

Right at this point, Roger disappeared forever, so I must assume
he was the culprit masquerading as The Nameless One and causing
great confusion.

> Do accept and trust Victor. He has worked so tediously and tried
> so hard to be faithful to his task. He felt so bad when instructions made
> him distress you a little. His concern for you is quite proper, dignified
> and genuine. Try to forget incidents he was forced to engender that

displeased you. It was all part of the training. It was necessary, but not welcomed by any of us.

If you can accept Victor and call him by his proper name, "Hweig," we would all rejoice that your distress and his genuine suffering have been salved by understanding and forgiveness.

Thank you.

We may go on, then, to better work than ever... *The Hidden One.*

And then, Hweig spoke:

Dearest girl, as I read over the first episode while you proofread, I can see why you trust me less than the others. I was rather horrible to you, wasn't I? I just didn't know that much about Earth women, but approached you as I did those in other places who responded to me willingly. I do apologize for my ignorance and misdeeds. I want more than ever to protect you now—Selket and I—and to write with you on these books. The Hidden One will give me his material and allow me to interpolate and make minor changes if you suggest them. We can work together as we were meant to do from the beginning. Only I goofed!

I am indeed the one appointed to work with you, but I don't know how to prove it. ...Ah yes, by behaving myself! You do put me on my good behavior, don't you?

Jamie and Amorto send their love. Always they keep track of what is going on. They would scramble me if I got too far out of line.

Your work will be needed very soon in a way you do not anticipate. I can hardly keep secrets at all, so I had better sign off. I so want to tell you!

All my love, dearest girl... *Hweig*

It is *so nice* to sign my own name at last!

April 14, 1979—9:35 P.M.
Hweig once again communicates:

Our little friends, the energy existences, have prematurely cued you in on some basic revelations. I prefer to do this my own way.

I have previously explained to you that we exist in more than three ways at the same time, due to time dimensions we will study later. It involves psychic time, not mechanistic time as you understand it.

We exist in our corporeal bodies, just as you do. We can extend into a separate psychic essence (as only a few Earth people can do when they have out-of-body experiences), and we can also extend ourselves in a semi-corporeal manifestation. Moreover, we are aware of the experiences, thoughts and activities of all three appearances simultaneously, for our mind is all one mind.

Our semi-corporeal self would look very real to you, and is very real, but of a less substantial corporeality than our physical body. You could, for example, walk right through us, or pierce us with any physical object without harm. ...Yes, you are right. That is why in the first pendulum episode, we told you to plunge the carving fork into your chest. We were only testing to see if you were in your corporeal or semi-corporeal form. ...No, we could not tell for sure without some test.

The UFO people from Planet X seldom visit Earth in their corporeal bodies. Why should they risk something happening to them when they can manifest themselves just as well in semi-corporeality and appear just as "real" to you and converse with you and visit with you? But if any of us want to react physically with you, we have to take you aboard a physical (or hardware) craft, as you call it.

This is the reason for the abductions, as formerly done. This is now forbidden without explicit consent of the individual. He must be told what is happening and agree to it in some manner. Such incidents are far more rare than contactees believe. Many who think they have had this physical experience have not. They have been comfortably situated in their own place, but taken psychically (or out-of-body). Others have been left exactly in their own situation and given a "hallucinatory" journey, in which everything would seem real—even corporeal— but was real only in a psychological sense. I wish I could think how to delineate that better. ...No, it is more real than a dream; it really happens.

We are trying to make contact with as many Earth people as we can, and in any way we can conceive of doing it. And we do have many subtle tricks, which we use without mercy. Accidents do happen, usually due to the panic of the contactee or some outside interference, but we never intentionally harm anyone. We strive always to protect in every conceivable way...*Hweig.*

The following is an excerpt from a much longer story about my own "abduction" experience, told by Hweig:

Amorto, Jamie and I came from our physical world aboard a UFO—one of the hardware kind—in your year 1940.

You, Ida, were traveling with your then husband and two younger men in a car on the desert between Indio and Blythe, California. As you came over the hill, the valley before you was crimson, as though with fire. You immediately thought of a forest fire, but could see no flames. Ida, please describe the incident from your own viewpoint.

It was in the middle of the night. The entire valley and the hills were ablaze with a powerful red glare that was too deep and consistent in color to have been caused by a forest fire. A great red ball sidled from behind a rock or hill. When I pointed this out to the men, they said it was the moonrise, but it appeared to be coming sideways. Because we were going to the right of the rock, I protested that it could not then appear to be coming out to the left.

The driver pulled to the left of the road and stopped. He nudged my husband and spoke to the other man in the back seat. The three of them went down the road a little way in front of the car and stood in the headlights talking earnestly for a long time. I could not imagine what they had to discuss for so long.

By the time they returned to the car, the red blaze had faded. As we drove down the road, a great white disk rose rapidly in the sky. "You see, it is only the full moon rising," they said.

I had grown up on the desert and had seen a hundred full moons rising, but this was something very strange. It did not seem quite right. "Why is it going so fast?" I asked. "It is just shooting up!"

"It only looks that way because we are going in the same direction."

"But it is going *up*," I protested. "We are *not* going up."

They all earnestly insisted it was the full moon until I gave up arguing. A little later, after a short nap, I saw a quarter moon.

"How can there be a quarter moon and a full moon in the same sky the same night?" I demanded.

"Clouds are hiding part of it. It's a full moon."

I looked and looked. I could not see a speck of cloud anywhere near the moon. Then I fell asleep again and forgot about it.

At that time, I had never heard of a UFO. More than twenty years later, when someone was talking about UFOs, I suddenly remembered the full moon and the quarter moon the same night, and wondered if the "full moon rising" might not have been a UFO....

All right, dear Ida, I will now take up the story again. You did, indeed, witness the UFO that brought Amorto, Jamie, and me to Earth. We firmly believe (but cannot at this time prove) that this meeting was somehow planned and engineered by your Guardian. Amorto told you this story once before, then we had to retract it as too premature.

Our craft that night had many sophisticated instruments. One of them was a scanner that allowed us to "see" into the car and its occupants. We observed the three men and yourself, and tried to speak telepathically to all of you. The two younger men in the car were too brainwashed to receive us. Your husband was not in mental accord. We made you forget what you heard and saw that night. This is what I said to you: *Do not be alarmed. We come as friends. Someday I shall speak to you again. Wait for me and expect me.*

I was told the knowledge of the contact was too soon, that I must erase it from your mind. So I did. But perhaps it remained in your subconscious, for you have said that, years later, you would find yourself looking from your window grieving and saying to yourself, "Why am I grieving? I've not lost anyone. I am perfectly happy. Why do I grieve?" We have decided between us that perhaps the memory of my words that night were somehow making their way into your emotions, if not thoughts. Or am I being too presumptuous to say you were grieving for me? But you have said that since our first telepathic contact with you in 1968, you have never had that feeling of grief again. Oh well, perhaps I do presume too much!

This is not the full story of this meeting by far, but the rest must wait to be uncovered another day...*Hweig.*

Later that month, I received a reply from Dr. Sprinkle:

April 23, 1979
The University of Wyoming
Division of Counseling & Testing
Box 3708, University Station
Laramie, Wyoming 82701

> *Dear Mrs. Kannenberg:*
> *Thank you very much for your letter of April 11, concerning your experience in receiving mental communications from UFO occupants. I appreciate your willingness to write to me, and to describe your writings and your willingness to share information.*
> *During the past ten years, I have been conducting a "Survey of psychic impressions of UFO phenomena." Approximately two hundred people have participated. They have completed a questionnaire regarding their interest in ESP and UFO experiences, plus a vocational interest inventory and personality inventories. After the inventories have been completed and returned, we have them scored and then provide the individual with information regarding the profile results, as well as how the results compare with those of other persons who have participated in the survey.*
> *If you are interested in participating in the survey, please let me know and I shall be glad to let you know how these messages compare with those from other persons who have been corresponding with me. For example, two women with whom I have been corresponding, for the past ten years, have an ongoing communication with intelligences who claim to be UFO occupants, and who provide a variety of messages on various topics, including science and spiritual development.*
> *If you have some kind of summary, or sample, of your messages, I should be most happy to receive a copy. Under separate cover, I am sending you a copy of two papers which I have written regarding investigation of UFO experiences.*
> *I hope the information is of interest to you. Thank you very much and best wishes to you.*
>
> *Sincerely,*
> *R. Leo Sprinkle, Ph.D.*

I corresponded with Dr. Sprinkle for some time:

April 26, 1979

> *Dear Dr. Sprinkle:*
> *Thank you for your letter of April 23rd, acknowledging mine of April 11, and seriously considering a summary review of my*

experiences. I have told no one about them except my husband—not even my mother, who lives with us. All she knows is I spend endless time in my room bending over a hot typewriter.

How does one summarize some 125,000 words!? But these are mostly my personal experiences of learning to work with these avowed UFO personalities and passing a few hundred of their interminable "tests, tests, tests."

Perhaps I can strike the heart of your professional interest best by simply sending you an outline of one of their projected books. Of course the most suitable title would be the "Outline for Interview in Psychic Research."[1]

At first they seemed to want me to learn to do psychic research from their viewpoint, but they soon realized my forté was writing rather than interviewing, and so decided to extend the outline into a book for anyone who was interested to use.

After I received your letter yesterday, they added material relevant to UFO experiences specifically.

I am awaiting the copies of your two papers most eagerly. I had hoped they would come today, but they did not. Also I am anticipating receiving your surveys and will be quite happy to participate in them.

I am not, by any means, 100% sure of the validity of this experience nor of the affirmations of my communicators. I range from an 80% believer on my "up" days to 60% on an average, but sometimes, on horrible days, my belief plummets to zero.

Thank you for all courtesies and consideration.

Sincerely yours,
Ida M. Kannenberg

May 10, 1979

Dear Ms. Kannenberg:

Thank you for your good letter of April 26th, and for the "Outline for Interviews in Psychic Research." I am very pleased to receive the summary, and I hope that you are willing to give me permission to quote from the statement you have provided me.

I am very interested in your writings, not only because of the possibility that they are received by you through mental communication from UFO personalities, but also because the information given you is very similar to observations other people have made who claim contact with UFO entities. Thus, it is al-

---

1. Note: "The Outline for Interview in Psychic Research" has been used as the basis of questions in a handbook to help other contactees cope with their experiences. "How to Come to Terms with Your UFO/Alien Encounter" has been presented in a separate book, *UFOs and the Psychic Factor*.

most as if there is some kind of overlap or confirmation of the
reliability of the experience. I am referring to "reliability" as an
indication that other people have similar claims; I do not know
how to deal with the question of "validity"—that is, whether the
information actually does come from UFO personalities.

I am sending you, under separate cover, the materials in the
"Survey of psychic impressions of UFO phenomena." When you
have completed and returned the materials, I shall be glad to
have them scored and to share with you information about your
profile results and how they compare with those of other persons who have participated in the survey.

I appreciate your willingness to share information with me,
and I look forward to further correspondence.

Best wishes to you.

Sincerely,
R. Leo Sprinkle, Ph.D.

May 10, 1979
Continued May 12, 1979

Dear Dr. Sprinkle:
Your papers arrived early last week, but I was held up in my
contemplation of them by participating in an antique show for
four days. Now that is over, and I can resume my studies.

Incidentally, the envelope arrived opened. I trust nothing
was lost in transit. It contained three papers: 1) "What are the
Implications of UFO Experiences?" 2) "ESP Literature and
Counseling Psychology," and 3) "Hypnotic Time Regression
Procedures in the Investigation of UFO Experiences."

Your paper on implications certainly encouraged me to believe more confidently in the reality of my experience, for every
single item offered under "Speculations" (on pages 11, 12, 13)
corresponds exactly with what I have been told. I was even
somewhat amused to see the word "bridge," for my informants
have told me on several occasions, "You are a bridge."

Since the leader of the communicators, Hweig, declares he
can, at all times, see through my eyes and hear through my
ears (discretion permitting), he also read your papers right
along with me. (Doesn't that sound crazy?) Hweig was very
pleased with your presentation in many ways. He would like to
offer a few comments: On page 11 of "Speculations," the word
"tensor" met with his objections. He understands others have
offered "tensor beam" as explaining one of their processes in
communication. He says, "This was a camouflage, necessary
for a time. It is not tensor beam, but laser beam." About
beams—tensor, laser, or otherwise—I personally know nothing, except that during my first episode ten years ago, I was

told on occasion that a laser beam was being used. I argued at that time that they must mean laser ray, because I had never heard of a laser beam. But they said then, "No, beam. It is our laser beam."

The phrase "beneficial to all mankind" is one my communicators use over and over.

I was interested to see under "Some Personal Writings" the mention of "Pendulum Technique," for it was through the use of a pendulum I was inducted into UFO experiences on both occasions. To use this without guidance, they tell me, is very, very dangerous, as other less friendly personalities can take over. (This happened to me on my first experience with disastrous results.)

At first Hweig said the shifting of study from contactees to the knowledge presented was a little premature, but after days of conference, he and his colleagues and advisors decided that the time is indeed at hand when this may very well be advisable.

You have referred to Koestler's concept of Janus-faced holons as a construct, referring to small energy fields that can behave either like "waves" or "particles." "Not fields," says Hweig, "but masses, which become fields only under very special circumstances and not of their own accord." There is a terrific lot he wants to say about these on another occasion.

Another comment he asks me to make: On page 10, you quote Orme as speaking of different levels of time. "Not levels," says Hweig, "But dimensions of time—far more complex than the idea of levels."

Ida M. Kannenberg

May 23, 1979

Dear Dr. Sprinkle:
I do not know if there is anything in that incident of a full moon rising on the California desert in 1940 that hypnotic regression could call forth. It would be interesting.

Also, another incident I don't have confidence enough in to report: During my first contact ten years ago (via pendulum), I distinctly heard a man's deep voice saying, "Do you remember the little girl under the lilac tree?" And someone else answered something. I told this much in the first episode, but not what has now been told me. I was told to recall playing under a lilac when I was about seven years old. That took a lot of memory digging, but I finally did remember when I saw a snapshot that had been taken that day. I was either sleeping or half asleep under the lilac when two men came up and spoke to me. I thought they looked like "very important business men" and they had a shiny black car parked a half block away. (That was about 1922 and almost all cars were black.) The larger and older of the two asked me where a certain person lived and I pointed out the house. They did not go to the house but got in their car and drove another way. Now my communicants tell me a message was given me, not vocally but "in my head," to study "about people and to learn to write well." I suppose hypnotic regression might recall something, if there is anything there.

Sincerely yours,
Ida M. Kannenberg

May 24, 1979—8:15 P.M.
How good it feels to finally talk to someone and share all of this information! Hopefully, many questions will be answered when we meet. For now, it's great to talk to someone who doesn't call me crazy.

Lessons of Atlantis

June 12, 1979—3:20 P.M.

Questions about their avowed manipulation of time and their relationship to Earth keep returning to my mind. If they can travel vast,
inconceivable distances in their "time ships," why do they choose to
come to Earth? There must be thousands upon thousands of planets
in the whole universe that would be hospitable to them, some perhaps
not now inhabited, but very adaptable to their living requirements. No
sweat, no fuss, just move in. So why Earth? Why now? There has to be
something more to it than their own willfulness or caprice.

Putting together bits of information and discounting at least 60%
of other information as deliberately misleading, I reach this conclusion:
Earth has a special meaning for UFO persons. "A common root stock,"
they once said.

Several times I have asked them if they come from our future. They
have not answered, but have dangled the idea of travel in a "time differential" before me. Could they just as easily project themselves into
their future? Could they be from our past? Do they come from Atlantis?

Forty-eight hours after recording this question in my journal, I received an answer:

June 14, 1979—3:45 P.M.

We cannot put an idea, fact, or information into your little head until
you have had a glimpse of it yourself. All right, dear girl, you have now
glimpsed our relationship to Atlantis. We told you previously that we,
the UFO people, came from two sources: a far planet, for which we
use the name Tea Elsta, and Source B. The latter planet, Source B,
is truly Atlantis.

There are ten facts to be told you now. Some repetitious, but all
are important. You may reveal these at your discretion:

1. We are never going to invade Earth in any fashion, but we do want
   to "come visiting." Plans are being worked out.
2. Our leaders are men from another planet. Don't worry, they look like
   you only more handsome.

3. We have work planets where UFOs are at present being repaired and new improvements are being worked out.

4. Our original ships, huge mother ships that we prefer to call space stations, were made in Atlantis many thousands of years ago and have come through the time differential. They have been in your stratosphere more than two hundred years. We work out of them. Our original home *is* Earth—Atlantis. Can you grasp this? ...Yes, only for two hundred years, more or less, have our space stations been hovering. They are now far, far out. Once they were closer, but a mishap or two convinced us that we, and you, would be much safer if we moved out into your stratosphere.

5. I am Chief Communicator. I have been on Earth this time for nearly forty years, helping prepare Earth persons for our coming. We have not decided how many of the smaller craft (provided by Planet X) to send. The number will depend on coming events. The huge (miles-long) space stations can never descend. Eventually they will be destructed.

6. Our Atlantean numbers are many, and of many varied kinds of beings, but we are nothing in number like the United States' population. Oh, my goodness, no!

7. Our landing place is not decided. It will definitely *not* be the U.S. or Russia. We fear this would create discord between you. It will probably be France, maybe Japan. Or maybe several places at once.

8. Your writing will help prepare the way for our coming. Just prior to that event, many contactees will come forward with their stories of contact. That is when the interviews in psychic research will become important.

9. Time goes so quickly on Earth. We have much to decide on and prepare. Be patient with us.

10. My part with you now is mostly to help you make necessary contacts and to keep you informed. Do not get balky! Time goes too swiftly to waste it arguing...*Hweig.*

...Yes, we did come from Atlantis to scout Earth conditions many times before choosing this era to emerge.

...All right, so I've told some phonies. Most of what I have said is true, though exaggerated sometimes. The wild ones are to throw you off the trail. You are not supposed to know so much so soon. What you have been given in writing is all true. Only the orals are sometimes distorted (or as you call them, "pretzelled") facts....

We chose to enter Earth's stratosphere at this time, your 1800 to 2000 A.D., as the time of greatest danger to losing our home planet altogether. We want to return to Earth. *It is home.* There are other reasons for choosing this exact time. They will be disclosed in a few days...*Hweig.*

In June, my test scores arrived:

June 18, 1979

Dear Mrs. Kannenberg:
Thank you very much for your willingness to complete and return the questionnaire and inventories of the "Survey of psychic impressions of UFO phenomena." Also, thank you for your additional comments on childhood experiences, which may be related to your UFO experiences and the "monitoring" of your development.
Enclosed are copies of profiles of the inventories that you completed. I hope that the related information, plus the copy of the summary of the results of other participants, is helpful to you.
Your scores on the Strong-Campbell Interest Inventory (SCII) indicate that your interests are primarily in the average range for scales on basic activities, with higher scores on the scales called adventure, art and writing, science, nature and agriculture. The pattern of likes and dislikes is similar to those women who are lawyers, reporters, librarians, artists, advertising executives, language interpreters, psychologists, speech pathologists, college professors. Thus, you score like those women who have both artistic or expressive and investigative or professional interests. (See back of profile for further information about the interpretation of the scores.) Also, the IE score (Introversion-Extraversion) indicates that you score like those women who prefer to work alone rather than working with others.
Your scores on the Adjective Check List (ACL) indicate that you score primarily in the average range with somewhat lower scores on the scales called Exhibition, Aggression, and Heterosexuality; somewhat higher on the scales called Favorable Items, Self Control, Achievement, Endurance, Order, and Counseling Readiness; and the highest score on the scale called Intraception. In general, these scores are similar to those of women who see themselves as more interested in the understanding of human behavior and less interested in close relationships with other persons. (See attached excerpts from the manual of the ACL for further information about interpretation of higher or lower scores on these scales.)
Your scores on the 16 Personality Factor Test (16PF) are primarily in the average range; you score like people who are seen as abstract in their intelligence; tender minded and sensitive; self-sufficient and resourceful; and shy or timid.
Your scores on the Minnesota Multiphasic Personality Inventory (not shown) indicate that your scores are in the average range (no neurotic or psychotic reactions) with a higher score than average on the scale called Social Introversion: thus, you

score like persons who see themselves as shy, reluctant to talk with strangers, and who feel uncomfortable in social situations.

All in all, your scores indicate that you are similar in your choices to those women who see themselves as interested in a wide variety of artistic and scientific activities, with personality patterns in the average range, except for the scores on scales which attempt to measure interest in social interactions or social gatherings.

I hope that these interpretations are helpful to you; if you have major comments or questions, you may wish to check with another psychologist or professional counselor. Or, if you have a specific question about my interpretation of scores, I shall be happy to reply to any question.

I appreciate your willingness to share information with me and to give your permission to use the information in my own work. Thank you so much for your time, energy and knowledge.

Peace and Light.

Sincerely,
R. Leo Sprinkle, Ph.D.

Of course, Hweig commented:

Your correspondence with us has now been underway for perhaps a year, and will hopefully bring you much relief from present doubts and worries concerning us "characters" (as you call us). To know that others are having similar experiences and to be assured of your own sanity must surely give you a great sense of security, and, we hope, will give you more confidence in yourself. You do so try not to impose your own thoughts and ideas on other people, but you will presently find it necessary to become more outspoken. You need not be aggressive to a point of aggravation to others—just quietly assured of the value of your own experience and thoughts. You must not be shy and retiring wherever your voice can add to the knowledge of our reality and good intentions. "Spread the word," dear girl—become our witness. Quietly do so, as becomes your nature. You have for too long remained a quiet mouse soaking in understanding and silently committing yourself to good works. It is time to let your light shine!...*Ole Uncle Hweig.*

June 21, 1979—10:16 A.M.

Yes, you have deduced our largest secret. We meant for you to do so and gave you many clues, but we did not expect you to come to it so soon. Now from various clues you can put together a whole story.

The first men who discovered outer space travel (in this cycle of humans) lived in what you call Lemuria. ...No, Mu was the entire known world at that time, and encompassed more than Lemuria proper. This was many thousands of years ago—fifty thousand and more...yes, even two hundred thousand! Lemuria was a continent in the Pacific Ocean that stretched far to the south. It is there true humans—the seed of Earth's present inhabitants—came first.

Many thousands of years ago (and there is a reason for not being more specific at this time) there was a splitting of the land masses, which marked out the continents of Europe, Asia, Africa, and the Americas much as they are today (taking into account the vast amount of erosion on their shores, plus natural catastrophes and human-made ones). Lemuria (in the Pacific) and Atlantis (in the Atlantic) were well defined at that time.

At the time of this splitting, space travel was in its infancy in Lemuria, which was then the most advanced of all civilizations. A certain faction, desiring to escape the land catastrophes they believed to be coming, called upon Superior Ones to aid them, as they had been aided in the past. This faction escaped in a number of created (not manufactured) craft and entered outer space in search of a new home. Eventually they found a planet paradise. It was much like Earth, except for atmospheric and climatic differences, which they soon learned to control. We call this planet Tea Elsta (their own name is something other than this).

Centuries later, after a great deal of division amongst themselves, the people of Tea Elsta devastated their new home by experimenting with something equivalent to atomic explosions. Some of them returned to Earth to join Atlantis (for Lemuria was no more). Most stayed hopefully on, trying to revitalize their planet. These are now the planners and builders of what you call UFOs. They are aided by other planetary people.

Many thousands of years later (but still many thousands of years before your time), the scientists of Atlantis not only dabbled in space travel, but had secretly conquered time travel. They were also masters of the natural science of psychic matters.

The scientists quarreled with the religions and the priests, who had subdued the people through distorted beliefs and practices. They wanted to retain the purity of the Old Religion, the knowledge of reality. They began to consider ways of escaping the power control of the priests. They did not want to leave Atlantis and planet Earth, as their forebearers had done nearly 30,000 years earlier. They began to experiment even more secretly with time travel, and, just as your people would do, began to experiment with the future. They tested age after age into the future. They wanted to see when, if ever again, people returned to the elemental beliefs in natural forces and beings, in the idea of one God who encompasses all, in the ideals of cooperation and productivity (rather than materialistic creeds of accumulation and misuse of their own natures). Throughout the ages, people have reported

UFO sightings of many kinds. These were due to the experimenters of Atlantis traveling into their future to find a time to emerge—hopefully into the civilizations of Earth.

They witnessed the second breaking up of continents. Atlantis became smaller islands, and they knew they must move soon. They witnessed the final breakup of the islands and their submergence into the sea. They tried to warn their people, but came into open conflict with the priests. The situation became intolerable.

They openly constructed what you call space stations in the name of experimentation. These were large enough and there were enough of them to evacuate a whole segment of their population. After nearly one hundred years of trials and attempts, they perfected space stations that could be used in time travel, but not in outer space travel. They provisioned the stations with great quantities of necessities. Foods were synthetic and in compressed form, so a great deal could be stored in small spaces. Water could be chemically produced, but not in the quantities that might be necessary. This became their great worry.

When all was ready, the scientific community (comprised of male, female, androgynies and a varied number of worker types), by various pretexts and devious planning, gathered into space stations and sent themselves into the future to the point of emergence. They chose to emerge late in the second millennium A. D., less than two centuries before humans of planet Earth would find themselves capable of again blowing themselves out of existence—*now*. They skipped the interim of history as though it were a day and night in their existence.

Now the Atlanteans desire to approach the current inhabitants of Earth and to find a place for themselves. To them, Earth is still home, even as it is yours, and now mine. They want to come in Peace, as friends. The planetary ones want to re-establish their world with flora and fauna from Earth (and perhaps with a few brave souls to revitalize their diminishing Earth-type populations).

You say this sounds like science fiction? My dear, a great deal is being put over as science fiction these days that is based on absolute fact!

We have capsulated the story as much as possible. ...Of course an enormous amount remains unexplained. Someday you shall write the whole story—but only after evidence verifying the reality of Atlantis has been discovered. Right now it would be wasted effort. You are yourself dubious...*Hweig*.

A short time later, Hweig elaborates:

All right, little goose, here it is: We of the craft of Atlantis left nearly 22,000 years ago, and have been orbiting Earth for less than 200 years. Our journey was through time and somewhat through space, as we told you. We do not go to reside on Tea Elsta, because our purposes are not identical to those of its inhabitants. We have the same *roots* as those on Tea Elsta, but we feel closer to Earth people. We want to return; the people of Tea Elsta do not. It is that simple. This is our personal and immediate purpose.

We act, however, within the framework of a larger plan, from which we receive instructions and to which we have to conform, accepting rejection of many of our own proposed plans. We can only move forward within the confines of the greater plan. We have told you we do not know the source of this plan any more than you do. Over centuries of experience, we found it best to obey. Even before we left Atlantis, we listened to the source and left Atlantis according to their instructions. Others in Atlantis would not listen and lost their lives.

Our part in the greater plan is to teach Earth people all we know about science, medicine, history, etc. That is a difficult task, since no one seems to listen.

Your job is to make them listen. Hence your writings! Is that not a clearly expressed and simple statement?

Go back to your reading...*Hweig.*

June 22, 1979

Dear Dr. Sprinkle:

I want to thank you for the interest you have taken in my experiences with UFO personalities—if indeed that is what they are!

(Hweig loves to tease more than anyone I know, but let him! In the midst of his teasing, I often catch onto things that he does not intend for me to know so soon. Or does he do that on purpose?)

Ida

P. S.

Here is something I have been meaning to tell you, but have forgotten to. I keep hounding Hweig to explain what he knows of animal mutilations, men in black, and Sasquatch. He has said repeatedly that:

1. *They personally have nothing to do with animal mutilations. They have been studying and observing, but the knowledge is too dangerous for me to know.*
2. *They have nothing to do with the "Men in Black" phenomenon either, but know the clique and believe them ridiculous and absurd.*
3. *Sasquatch are indeed their creatures, let out upon Earth to test energies emitted from the ground or Earth itself. We do not recognize these energies although they work upon us, sometimes quite detrimentally. Sasquatch is not ape, not human, but a biological creation about halfway between the two, with no biological connection to either. It is not a "missing link." The*

creatures are mild, unless startled when with their
young. They have been unloaded here for years, and
now have acclimated and oriented themselves. They
do have offspring, but rarely. They are monitored by
pictorial telepathy.

July 3, 1979—6:30 P.M.

We have been consulting those in authority to see how to advise
you in your research. We cannot hand you facts on a silver platter—
one, two, three—because that would not develop your own abilities,
but when you have come into knowledge of an area of fact, we can
extend and help clarify.

By all means, go to Virginia Beach. Edgar Cayce will, someday
before long, have the credit he deserves for his work on reincarnation
which involves much about Atlantis. In the library there, you will find
much useful information on Atlantis that has never been published, as
no one had a clue to its significance. Two or three days should be suf-
ficient. You will be guided or aided to find information or clues that will
take you to the Smithsonian in Washington D.C. There you will know
what to look for and what to do about it. However, the records at Virginia Beach will reveal all you really need to know if you do not want
to continue on to Washington. If problems of any kind arise, just depend on Ole Uncle Hweig. ...Yes, you will be able to find the books
you want. Be sure to make a list!...*Hweig.*

It does not seem necessary here to go into details of my excursions
to Virginia Beach and Washington D.C. Much information was gained
in the Cayce library on Atlantis and on psychic matters in general. The
history department at the Smithsonian was more interesting to my tag-
along guests than to me. The Sumerian and Mayan artifacts that I
wanted to see were either no longer on exhibit or locked up for the summer.

July 29, 1979—1:30 A.M.

This morning let us speak a little more of ourselves. We did con-
quer time travel and can go forward or backward in time, but simply
have no reason to go back. We wanted to avoid the in-between years
(when civilization again sank so low and for so long), and to escape
the birth and re-birth cycles, and so we skipped over, in one time leap,
from thence to the present, when so called civilization, has once more
attained something of the old ideas and projects. We wanted to re-en-
ter Earth at "progress time," just a little before where we left off, and
have done so.

Forget the speed of light and all the factors thereof. There are easier and better ways of beating time as you know it—better in that we
can go forward or back, or even sideways (which you cannot understand, but will before long). ...Ah, you do have a glimpse that what I
mean is possible, but you do not know how to explain it. Good! That

gives me an open door to talk about it, and before long I shall put something together to report on this.

Ezekiel, the Sumerians, the Mayans and others all were visited by Atlantean time explorers. Many more places were visited, some of which you will uncover for yourself.

In the days of Atlantis, manifold opportunities arose to work with the nature spirits, which modern man and woman have tossed out the window. (Your American Indians are not so wasteful of their opportunities. They know better. Study the history of the Iroquois and the Hopis and listen with your "inner ear"—your inner understanding.) More than this, we were able to call upon the creative forces from other realms of being to advise, counsel and direct us in our undertakings.

Your species is now entering that same point in its own development. The spoon-bending kids of this generation are only verifying and calling attention to the fact that there are greater powers to be utilized by humanity than are recognized or appreciated at present.

Listening with your "inner ear" does not mean listening to the small, wee voice of conscience. Some reach this inner aid by contemplation or meditation (good methods, though comparatively feeble ones). Some use intense prayer. Some use a fierce inner concentration (which is your way), although it is seldom so rapid of results! ...Yes, even I, an adoptive Atlantean, speak to you in this manner. It is a very high form of telepathy that your researchers will never discover by playing with cards and such tedious methods.

Now for myself: I am, indeed, an Earth person born of an Earth mother and a fly-by-night Atlantean father. My mother died when I was a very small boy and I was taken (by the UFOs) to be reared and educated as one of them. So I am half-Atlantean and half-modern "Earthian." It is sometimes very confusing, even to me, exactly where my sympathies and future lie. In 1940, I came to Earth (via UFO) to study in your universities and learn all I could of the psychological and allied sciences so that I might interpret modern Earth persons to the Atlanteans (for, even as they seem so alien to you, you are alien to them).

I was told time and again that my correspondence with you had to end, but I always howled, begged and demanded, until I was allowed to start our conversations and writing again. Even now I am not supposed to engage in personal talk.

My educational years on Earth were strange indeed. In my own body, I remained secluded; I attended your universities and classes out-of-body. You will find no records of attendance naming me either Hweig or Victor Azimorov (my Earth name).

I had permission to attend classes only (not to enter into anyone's home or place of business to investigate), until I was given direction to cultivate you and direct your abilities into specified lines. I was indeed clumsy and most inadequate to the task, but, thanks to your divine patience and equally divine tongue-lashings, I have learned at least what *not* to do—and all of this has been grist for the mill of the home people (the Atlanteans of the space stations).

It does read like a fantasy, I know. Proof will come shortly. Not evidence, dear girl, *proof.*

I want to deliver a second fact today. We are not permitted to reveal all I would like to tell you, but I can say this much: You are without any knowledge of a type of personality that has come into the world at this time specifically to interact with us and help bring our two peoples together. But you are not altogether ignorant of what this is all about. It all lies dormant in your mind, awaiting the proper time or moments to bring forth each fact that is needed as it is needed. Do not fret about not being able to find answers. You already know them. As you seek and try to write your understanding of things, the facts will seep out to your attention. Has it not frequently done so already? (However you allow yourself to become confused by listening to everybody else! Try to distinguish opinion from fact and you will be all right.)

We can see into the future only as probabilities, although in some rare authorized instances we can see events. Any situation depends on opportunities offered and free will to accept. Therefore, we must patiently wait until a choice has been made, and only then can we look into the outcome. Otherwise, we would warn or instruct, thus, interfering with freedom of choice. ...Yes, that is splitting hairs pretty fine, but it is the way we must operate. We can give advice only according to probabilities...yes, hope only. The creative forces can instruct, advise and direct, but we are only human like you. We cannot usurp the responsibility of your choices. I have prognosticated things for you only on the basis of probabilities. I had hoped to bolster your enthusiasm, courage and good spirits, but I have so often fallen flat on my face (and discouraged you all the more in the long run) that I shall refrain from this sort of thing in the future. I do not have that kind of control over things. Sometimes I can engender aid from those who can foresee (or even manipulate) events to some extent, but that again is meddling where I have no real right to do so. And my idea of what is right for you may not be all that wise (as I have discovered to my regret and your discomfort).

It is too bad that I was not released to explain all this sooner. It would have "saved a lot of wonder and worry," as you point out...*Hweig.*

August 26, 1979—8:42 P.M.

No work is wasted. Believe me. The cause is as stated. So are we. But there will be no more tests and experiments. You are too nice and sweet to continue to harass.

We are exactly as we have told you. But no more impelled trips, studies, etc. We give or bring to you all you will need.

...Of course you don't really understand, little goose. ...Of course this is Mordalla, not Hweig. Hweig will not be in contact for some time. I am Mordalla of the UFOs. Yes, Hweig and I got very confused before in your mind. I said I would come back and give you information on Atlantis and UFOs, didn't I? Here I am. I have since learned to talk direct-ly, without an interpreter.

The time has come for more surface contacts and activities. ...Do not worry, the statue can be mended. (I had knocked over my little statue of Selket, the goddess who guarded the inner shrine of King Tut, while cleaning. Her hand had popped off and I had cried over the accident.)

Your work, dedication, loyalty and helpfulness will not go unno-ticed where it counts. Be at ease, girl. All is very well...*Mordalla.*

Later, Mordalla spoke again:

Do not be alarmed by this short interval of non-action. Take anoth-er week to get your work room in order. It looks very efficient. Get pa-pers filed and everything arranged for quick finding and recovery when needed. You are going to be busy with a variety of tasks.

The period of psychological testing and experiment are truly over. You worked through a great many psychological problems that we made up for you. We know you extremely well now. Your loyalty, de-termination and work habits were clearly demonstrated, as were many hidden results that you were not aware of. Now we go on to real work. There is valuable information to be given and gained, and you can help as you have always done.

I am truly your old acquaintance from Planet X, but of a little differ-ent order of being than the human types of that planet (as I was only a visitor there). Now I am a visitor on Earth. I dwell, for the present, in the same order as The Hidden One (who is truly your Guardian). We are located not too many miles from you, but are not letting ourselves become visible to Earth persons yet. (I said that wrongly.) We are vis-ible, but stay within our quarters and do not go out into public view. A matter of discretion only. Mordalla is my true name (or one of them) as closely as it may be given in your speech.

There are many tiers and hierarchies of personalities other than human. I am of a different type than you. My corporeality is less dense and of a higher vibration.

You feel the constant physical vibration throughout your body be-cause the vibratory rate has been increased much beyond normal. This assists in the telepathic contact and allows us to see through your eyes and hear through your ears. The Hidden One, Hweig and I are the only three in contact with you who can do this by our own abilities. Others can do it through instruments. Scientists in your world are working on projects that will allow them to do this.

We, too, are happy to be able to carry on a straightforward and open communication at last. The camouflage was equally tedious to us.

You may feel you have indeed "graduated" (as your charming Dr. Sprinkle called it), and with great honors! The hypothesis you sent him you may consider as your thesis—not the writing itself, but the fact that you were willing to let someone else in on your ideas. Had you clutched it to your ample bosom and declared it all your own valuable secret, you would have flunked the whole course.

We could not tell you before what was really going on and why, because doing so would have influenced your decisions, even if you did not mean for it to do so. But in all innocence and sincerity, you always made the right and honorable decision.

Hweig has greatly overworked himself and will need a long rest. We tried many times to get him to take time off, but he was quite jealous of the work he was doing and would not let anyone else take over. We also suspect he was equally jealous of someone else working with you, as a personal thing.

He is a vastly changed person after working with you this long. He did have a drinking problem. Strangely, you perceived this, and asked him to stop drinking when he was working with you. He stopped altogether. "If it is so obvious that she can detect it through all our barriers, it is too much," he said. He was truly quite a lusty and earthy character. Thanks to your gentle and refining influence, he has become a more calm and collected gentleman. The success of his work with you (from our viewpoint) has elevated his scientific status. His social status (as you would call it) was actually tops, as he is regarded of the ruling class. In the Atlantean communities, the scientists—those who have knowledge—are the rulers.

The little sketch he gave you of the manner and reasons of the time escape was all very true, but must be greatly extended to illuminate many questions that are bound to arise.

I eagerly await the time that we can start on a fuller exposition of your hypothesis. Much evidence will be found before it is feasible to continue. Your present guide, friend, and companion...*Mordalla.*

The following is an excerpt from one of my letters to Leo Sprinkle:

I have just come across the excellent book by Warren Smith titled, The Book of Encounters....[1] *Nice to meet you amongst its pages.... Smith's book has some interesting paragraphs relative to my own thoughts. Raymond Shearer (page 126) said of the UFO crewmen: "Some pictures of the old Egyptian gods have a distinct likeness to these people."*

According to various stories, pre-dynasty Egypt had an influx of Atlanteans, who were superior in every form of culture to the other amalgamating people, and, in much later days, these early leaders became legendary as gods and goddesses. So, of course, the gods of Egypt (the Atlanteans) would be similar to Shearer's description of the UFO personalities.

1. Smith, Warren. *The Book of Encounters.* New York: Kensington Publications Co., 1976.

Herb Schirmer (same book, page 112) said of writing he saw on board the craft: "This stuff was more like symbols, like the stuff you see in the movies about Egypt."

Hweig told me, long ago, that Egyptian writing derived from Atlantean. (I suppose the Rosicrucian symbols that look Egyptian are actually derived from earlier Atlantean forms.)

The serpent with wings—"the flying serpent of olden days"—that Schirmer drew (page 112) would definitely be Atlantean.

Now about Hweig: He is truly gone, and I am desolate. For two months, others have been telling me he was worn out, depleted, and simply had to take a long rest. It seems he battled about this until he helped me get the hypothesis stuck together and mailed. That was the reason for his agony in getting a copy mailed to you. Perhaps he feared the others would let me change my mind about it.

Wordy, as always,
Ida

P.S. I was just asking myself what the flying serpent symbol would actually mean in the old Atlantean usage, and the words came into my mind "profane knowledge." So I said, "Hey, Mordalla, is that what the symbol means?" "Actually," he says, "Secular knowledge in opposition to knowledge of spiritual things. Those who wear the flying serpent symbol are scientists, not priests."

P.P.S. I just asked Mordalla if scientists wore the symbol of the flying serpent, what would the priests wear? The answer was so obvious that I should have thought of it. "The Phoenix, my dear," says Mordalla.

Thus Speaketh Hwieg

September 1, 1979—11:10 A.M.
Mordalla begins a discussion (and I continue to wonder where Hwieg is):

> People seem to be thinking about UFOs with everything except common sense.
>
> If, as the book you were reading suggests, an alien horde were to blast all the people off the Earth (and otherwise, what would we do with them?) the planet would not be fit to live on for centuries (as the people of Tea Elsta discovered, for they had to go underground under very primitive living arrangements until the surface of their planet became habitable again).
>
> We do *not* want the problems of feeding, housing, clothing and training all those millions upon millions of people! We have far better workers already. And far better planets to go to if we wish, or if it becomes necessary to go elsewhere. Earth and its people are no prize whatever!
>
> Again and again, your smart people have calculated the possible number of habitable planets in your own universe, even in your own galaxy. Thousands upon thousands of choices, many as pristine as Eden, are available to us. Why should we want the polluted overpopulated planet Earth?
>
> You have already said this, dear Ida, in different ways, but I am irritated when I read such drivel and must speak for myself. You will find, my dear, that I am a bit more peevish than our beloved Hweig. But I do hope we can work together with the same camaraderie that you two were able to establish.
>
> We know his leaving has grieved you truly. He understood all that you thought and felt. His legacy to you is the contents of his thoughts relative to the work you undertook together. He spent night after night reciting things into your head while you were slept. He was urging you to go to sleep earlier of late so he could get everything told before he had to leave. ...No, your brain, your mind, your own thinking has not been tampered with. Only your information has been vastly enriched.

When you need a thought or fact, it will come forth. He knew well how to "time" all this (not an actual hour or mechanical timing, but the timing of need).

Now go on with your own work. I simply had to splutter a little. Who wants your darned old polluted population anyhow?...*Mordalla.*

On September 4th, 1979, I began again corresponding with Dr. Sprinkle:

Dear Leo,

Just a note to go on record.

I wrote to my ex-husband to see if he could tell me the month and year of our trip to Arizona (which resulted in the observation of the full moonrise on the California desert that the UFO boys are telling me now was a UFO). He said it was December, 1940.

A lengthy discussion with my mother...established that we were at my sister's for Christmas dinner. That means the moonrise took place about December 22nd or 23rd, but I cannot recall between what hours in the night.

Good enough. So I wrote to the weather bureau in San Diego to see if they would dig up for me the record of when there was a full moonrise in the month of December, 1940, and the time it would be observable in the Indio-Blythe area. I just wanted to get this on record before I received an answer from the Bureau.

Now I chew on my fingernails and wait. I suppose there are other ways I could go about getting this information, but this will do.

I am not at all convinced that there was a UFO instead of a moonrise. Or they could have coincided. However, if there was a quarter moon at that time—as I thought I saw later—then I am indeed due for hypnotic regression.

This was the only way I could think of to get a lead on the "true-false" aspect of the question.

Ida

Mordalla's communications did not last long. He collapsed in the middle of my sofa one night. (I am being facetious. I must not make jokes or you will not know when to depend on what I am saying.) I was lying on the sofa and began to shake and tremble so violently that I mentally shrieked at them to stop doing whatever they were doing. After a few seconds, someone said Mordalla had sort of "collapsed" and I would not be hearing from him again. All sorts of explanations and suspicions followed, much too lengthy to go into, but I now have Amorto—a tried and trusted old friend—on tap. I am glad, for he is the one I have always trusted the most. He has always shown deep concern and a very rare understanding of my feelings.

They tell me my future writing career for them has been chopped off in midstream. I will finish the book on psychic research. All the enclosed material will be included, and I will eventually work out the hypothesis, which is buried temporarily. The other writing tasks have already been given over to someone else.

Actually, my loves are archeology, anthropology and psychology, not psychic phenomena. I can do map dowsing and they want me to practice that. Also, I can do decoding if it's not too difficult, but I'm no whiz. So my future work with them seems to be intended to go along some of these paths instead of just taking down dictation. They say they have others quite as capable of dictation as myself, and in fact, they prefer to use them. They do not argue so much. I want to understand everything as I write it or I do not want to write it at all. I am too suspicious!

I have not yet heard from the weather bureau in San Diego relative to what time of December 1940 the full moon rose. But my husband suggested I call the local library reference department, and, sure enough, the lady called back a half hour later with the information.

A subsequent letter from my ex-husband indicates our encounter with the "full moon" took place on December 22nd-23rd, 1940.

The librarian gave me the phases of the moon for that month as follows:

First quarter: December 6th, 8:01 A.M. PST
Full moon: December 14th, 11:38 A.M.
Last quarter: December 21st, 5:45 P.M.
New moon: December 28th, 12:56 P.M.

Apparently, that quarter moon I saw in the sky the same night we also saw the "full moon rising" was the one that was really supposed to be there. My solar plexus is getting curly again! Maybe these fellows are *not* lying when they tell me I was on a UFO that night!

September 15, 1979

Dear Leo:

I promise a comparatively short letter this morning—not more than three pages, single spaced!

I just received an answer from my letter to the weather bureau in San Diego.... [It] corroborates the librarian's information that the full moon rose on December 14th, 1940. There has been some conflict in memories as to the exact time we arrived in Phoenix.... My ex-husband's letter established the time and date as 4:30 in the morning of December 23rd. Our driving time from Bremerton, Washington, to Phoenix was 52 hours (he always remembers driving times). We lost 3 hours in L.A. to car trouble and spent other time eating and getting gas. There is an hour's difference in Phoenix and Pacific time; there would have been no daylight savings that time of year. We left Bremerton on December 21st (if we arrived December 23rd) at the stroke of midnight going into the 21st. We did not stop except for eating meals and getting gas. We did not waste an instant of time, for the driver had to be in Austin, Texas, to get married (I believe on Christmas eve). He was a very quiet, almost stern, fellow, but a seemingly pleasant person. I remember how rigid he seemed to be for one so young.

My ex-husband says he does not remember getting out of the car. This is strange, as he has an excellent memory for just about everything—much better than mine. He remembers someone saying something about "the moon not being in the right place."

I think the driver had some knowledge of such strange objects, knew about what it could be, and took the men down the road in the headlights to talk about it. It is strange that my ex-husband does not remember. I asked him later what they had talked about. He gnawed on his knuckle and said, "Oh, nothing important." When he gnawed on his knuckle it meant he was studying something. The subject dropped there and I forgot all about it. Perhaps he did not remember and thought I was dreaming (or nuts).

I can't think of any way to get more information on this, except to hypnotize my ex-husband to see if he could dig up what was discussed. I am not sure he would go for that. I haven't talked to him in 23 years. (Silence is golden.) The only other way would be...to check records of UFO reports to see if the driver turned in a report to the Air Force.... He might have been the type to make a report. He was a very serious person. I could not do this myself.

I do remember looking at my watch soon after we stopped the car. The men went down the road and stood in the headlights talking.... I got impatient and tired of waiting for them. I kept looking at my watch and thinking "What can be taking them so long?" or something to that effect. Finally, I wrapped the blanket around me and thought I'd snooze a little until they got back. After awhile, I stirred, sat up, and looked at my watch again and thought, "My God, they can't have been talking this long!" But there they still were in the headlights. I tried to look at my watch better, as the light wasn't very good, and decided I had looked at it wrong the first time. At least, this is the routine of thought I remember now.

Of course, my UFO buddies tell me the men were frozen in the road and I was taken into the UFO because I was the only one they could reach telepathically. This part I am still very dubious about. I do have a feeling relative to this about walking up an incline covered with small loose pebbles and my left foot slipping back so that I lurched forward a bit. And in my mind there is a picture of myself approaching an object or room with a lighted interior, and when I slipped on the path, someone touched my left elbow to steady me. This could be all plain fantasy (or even fancy fantasy). I have thought too much about it. It is beginning to bug me. I wish I could find out the truth.

Now to go off on a new tangent:

Many times as I have gone through various events imposed by our UFO personages (as they call themselves), I have kept saying to myself, "But this is like an initiation. This must be some kind of an initiation," etc. There were solemn question-and-answer sessions, vows, strange gestures (I did not understand why I was making them), the bestowal of names and talismans. I always had the feeling of being initiated into something—and, indeed, they used such words as "neophytes" and "adepts" quite freely.

But the Betty Andreasson affair is certainly an initiation from start to finish!

I am beginning to think my entire life has been an initiation, or one initiation after another. Strange, strange, strange.

Thanks for listening. No one else does.

Sincerely,
Ida

September 23, 1979

Dear Leo,

Amorto has promised to give us some more symbol interpretations relative to the Betty Andreasson affair. He asks, however, permission to address you directly. Please meet my good friend, Amorto.

Dear Dr. Leo,

We have been aware of your work and dedication to UFO phenomena for quite some time. We have not made any direct contact and have not monitored you directly for fear of interfering with your scientific status as an objective data seeker. However, to express our appreciation of your dedication and concern in this project, we wish to make you a 49th birthday present through our mutual friend, Ida. You are the one who will know best how to use this to the benefit of others. It is to be a complete revelation of the symbology and meanings of the Betty Andreasson episode.... I, myself, had some part in the planning of the episode, and my colleagues have asked me to interpret it for them (and for you and whomever you believe

*would benefit by the knowledge). The material is completely
yours, with our respects and kindest regards.[1]*

Amorto and Brethren.

*P.S. Ida keeps asking me if the Great White Brotherhood[2]
has gone coed. Sometimes her jokes are more profound than
she knows.*

A.

October 1, 1979

Dear Ida:
      *Thank you so much for that marvelous birthday gift (the
message from Amorto). Please give him my "thank you" and my
best wishes for his continued efforts on the behalf of all humankind.*
      *I plan to send copies of the statement to selected UFO inves-
tigators, in hopes that they will consider the information as an
indication of Betty Andreasson's "initiation" and as an indication of how Earth scientists can begin to appreciate the com-
plexity of UFO phenomena. Someday, I hope that more
scientists will join us. Meanwhile, I am pleased with the information you gave me; thank you!*
      *Enclosed is a copy of a notice about the London UFO confer-
ence—complete with cartoon!*
      *Also enclosed is a survey.... I understand that 40,000 per-
sons are being asked to participate. Results will be interesting.*
      *Please inform Amorto that I value his knowledge and your
good efforts. I welcome further opportunities to share information; I look forward to the day when we can communicate more
directly.*

Peace and Light,
Leo

October 11, 1979

Dear Leo,
      *Something quite good (I think) has happened that makes me
wonder just how much and what kinds of powers these self-
styled UFO characters have.*

_____

1. For more information, please see Ida M. Kannenberg's *UFOs and the Psychic Factor*
    (Wild Flower Press, 1992), pp. 51-54.
2. A secret metaphysical organization composed of workers of "white light" that has
    existed on Earth for centuries.—Ida

1. *Twenty years ago, I fell in love with a tall, rambling Victorian house. It was for sale and I walked back and forth in front of it, not daring to go in, and wringing my heart (if not my hands) that I could not possibly afford it.*

2. *For four or five months, my husband, Bill, has been begging me, every time he stumbles over something, "Please take some time to hunt for a bigger house for us."*

3. *Hweig and Amorto have both assured me that, within three months, I would have a fine big house—big enough even to please Bill.*

4. *About two weeks ago, I recalled that Victorian thing and tried to remember in what town I had seen it. I have lived so many places.*

5. *Last week I had an errand in a little town abut 18 miles from our city.*

6. *I had to wait a half hour for my return bus and suddenly remembered "the house" and wondered if it could have been in this town, and if so, where was it?*

7. *I began to amble rather absently down 2nd Avenue.*

8. *There was the house—completely and beautifully restored and painted the exact color, soft buffy-yellow, that I had always imagined it should be. It was complete with white shutters everywhere.*

9. *There was a "For Sale" sign.*

10. *When I told Bill I had found our house and had been in love with it for more than twenty years, he said, "Why didn't you buy it?"*

11. *"At that price?" I said. "Do be funny!"*

12. *"Haggle," he said. "Or, since you don't like to haggle, I will."*

13. *Next day, he signed a contract to pay a lesser amount in three installments over a one-year period.*

So now we are busy as a swarm of bees, liquidating everything in sight. We have accumulated so much junk and stuffed it away into corners—now we are hauling it out into the light of day, viewing it with a mercenary eye and finding that (with inflated prices and appreciating values) we are somewhat richer than we thought. I am in a state of shock.

Ida

December 30, 1979

Dear Leo:
  Your trip to Brazil sounded like a whirlwind success. Where next? (Our ambassador-at-large.)
  My dear Hweig has returned (as I was sure he would, the wonderful, naughty rascal). I wonder who he really is? Is he—as he describes himself—a dwarf with four digits on each hand and foot; androgynous; half Russian (mother) and half fly-by-night UFO personality (father); a couple of centuries old, periodically rejuvenated? What a fantasy!
  The first of February, Bill and I and my daughter, Lee, are going to Hawaii for about three weeks. We are taking Lee to drive for us, since neither Bill nor I do anymore.
  For a year, Hweig has been telling me to get acquainted with my camera. I wonder if Hawaii—especially Maui—has some special import? It seems to, from the various things Hweig says. We will fly from San Francisco to Hilo, drive to Kona, fly to Kauai, then spend five days on Maui at Lahaina, and six or seven in Honolulu. Hweig has fooled me so many times (building up expectations that do not materialize) that I am not putting much stock in his insistence on Maui. I'll wait and see.

Ida

January 19, 1980

Dear Leo:
  Always glad to take part in any study, research, survey or whatever connected with our UFO experiences. The more we cooperate and share information, the sooner someone will come up with some concrete answers to the many questions.
  Re: Conference in May, count me in by all means. Looking forward to it.
  Hweig has promised some good pictures in Hawaii, but he giggles when I try to pin him down. I never expect too much of his promises because, if they happen, they are so behind-the-scene that I don't even know it. I'll just wait and see, I guess.
  You called the book on synchronicity to my attention (Incredible Coincidence)[3] I have not read it all, but I read a resumé in the January issue of Fate *magazine.*
  I mentioned in a letter to another contactee that it seemed the "ufolk" (if I may adopt your splendid word) were working against the pressures of time on one hand (trying to do too much, too fast), but on the other hand time, seemed to have a different significance to them (or clock and calendar time

---

3. Vaughan, Alan. *Incredible Coincidence.* Philadelphia, PA: Lippincott & Co., 1979.

seemed to have no significance at all!). So, of course! The UFO people do not work within the framework of time. They ignore any such mechanical measurements of time. They work within the framework of synchronous events (not quite the same idea as synchronicity that Jung and Vaughan gave it).

Their work is carried out by very compartmentalized factions, but any delay on the part of any one faction creates a hold-back for all. They have to try to "keep up" and sometimes "wait for the others." This is why sometimes I have felt that they were rather ruthless and short-tempered and they were working against time. It is not time, but synchronicity.

The only way so vast a program would work is by synchronous events, not an actual time schedule. The ufolk or their instructors have not invented synchronicity, they simply use this very natural process to order their events and the coming together of events. Therefore, Hweig is completely unable to tell me exactly when one of his promises or statements will "come true." It depends on the synchronicity of the overall picture.

1. *Various factions do not necessarily know any other faction or any other part of the larger plan than their own. They obey instructions and may not even know the final purpose of the instruction they are carrying out. They have faith and obey, and possibly have only enough information to allow a necessary flexibility.*
2. *The workers come from widely separated places, and are various types of beings. Some as alien to each other as they are to us.*
3. *Within each faction, each personage is as unique and individual as we are. They have various personalities and manners of doing things (except for the clones, who are semi-programmed.)*
4. *There is an infinite number of jobs to be done, and each with its own requirements.*
5. *If several factions have personal contacts, or simply parallel work or their work is contingent upon that of a "neighboring" faction, there can be conflicts. These are usually quite minor, but are delaying nonetheless. Various factions do not always agree with each other. ("...the understatement of the year," says Hweig.)*
6. *While all work toward a common end with a common purpose, there are discords. Heads bang into heads. Jealousy is at a minimum, and violence among themselves is unheard of, but there are still variances of viewpoints and judgments.*
7. *Any delaying fumble within one faction creates a chain hold-back everywhere resulting in nervous impatience and flaring tempers—even as with you and me—or at least me! (And can Hweig cuss!)*

8. *The ufolk are extremely sensitive personages. They are also extremely emotional, except for the uncommon few of most rigid self-discipline, who convey calm and composure, even under great stress.*

I believe the above factors explain not only the need for synchronicity, but also some of the seeming discrepancies in UFO contact reports. It is all so vast a project, and the variances are equally vast.

Ida

January 27, 1980

Dear Leo:
  Obviously the ufolk are infiltrating and working from the inside out, both through invasion of our minds and also in person (those who can pass as our own and those who have been taken from Earth, re-programmed or brainwashed and returned to do their appointed jobs here.
  I cannot believe that this invasion, mental or otherwise, is all sweetness and light. I see too much evidence of hardship and rough handling among the contactees and have experienced it myself. "Just teasing" or "testing" or "experimenting" is not always the full story. And to accept them only as benefactors of humankind is to make ourselves vulnerable to perhaps unpleasant surprises. We must remember that they, too, are "humankind," and the benefits may be their own.
  I was told at one time to take what they told me in writing as approximate truth (as far as bits and pieces can be truth), but to disregard most of what had been said verbally, as it was only told for the purpose of keeping me distracted, to keep me from putting two and two together. Above all, I was told anything to keep me interested and working with them. For this reason, they also make romantic overtures to Earth women who are particularly lonely or alone, or who feel themselves misunderstood or neglected by their mates. I believe this is why, if they do not instigate, they at least do not help alleviate the contactee-mate problems.
  Only uncommon common sense can keep one from falling into their snare of sympathy and understanding. They use every trick in the book to keep us working with them and for them. They promise the coming true of every hope, dream and ambition we ever had. Of course, they want me to think I shall take pictures of UFOs in Hawaii. I cannot believe it. I have always thought what a smashing triumph something like that would be. But I don't really anticipate anything like that at all. Should I be wrong and it does occur (and I have more than one roll of

film), I shall send one unfinished photo to you to have developed through official and controlled channels. I would be grateful for such an opportunity, but completely astounded.

They can also put us under a kind of "spell" (to use the old witchcraft word), that freezes our resistance. As long as they are speaking, we believe utterly what they are saying. Only later, when they let go, can we begin to reason and question. They have too much psychological power and use it too subtly for me to have absolute confidence in them.

Now to lay aside my negative thoughts. What if everything they seem to want us to believe—that they are here to aid in the Great Plan, to rejuvenate Earth and its inhabitants—is true? They are no more superhuman than we are. They are not miracle workers. They are definitely perturbed when contactees seem to expect miracles from them. Over and over they say, "We have great psychic powers, which are not supernatural and not spiritual, but entirely natural, and which are latent (if not dead) among your own people. We are technologically and scientifically considerably advanced over your people, but that is all. We pray just as hard for justice and peace and good will as you do!" Now they sound warm and chummy and altogether believable, and one longs to cooperate with them for such ends, which leaves me somewhere astride a fence.

For this reason, I feel compelled to go ahead and correspond with other contactees, learning all I can from them concerning their contacts and trying to find some common factors, some directing principle underlying the whole phenomenon.

You shall now have a vacation, for I am not going to write again until I return from Hawaii (Feb. 18th), unless something startling occurs. Should anything terribly exciting happen in Hawaii, I will phone. You'll be the first to know!

Ida

February 10, 1980—10:08 A.M.

I am so sorry I deliberately misled you on the pictures on Maui. I am foolish to take such chances with your good will. Of course, you are annoyed and a little righteously angry!

There is a very good reason for you to come to Hawaii and to have a good camera. And that reason will show itself before long. Please believe that and forgive me and do not doubt who, what, why we are, for we are just what we have told you.

Bear with us a few days longer—no promises, no intent to mislead or offer false hope. Just bear with us a little longer and all will become evident. My word of honor...*Hweig.*

February 20, 1980—5:15 P.M.

We are going to give you a chance to get caught up on your business, bookkeeping and housework before getting too involved in other studies and writing. But to begin:

We are exactly as we previously described to you—human beings, much like yourselves, connected with the UFO phenomenon.

I, Hweig, was born of a Russian mother and a UFO Atlantean time-traveler father. I have characteristics of both types of humans and abilities (inherited from my father) that present Earth people have lost, the main one being psychic mastership.

Amorto and Jamie were both born on the huge timeships of Atlantean parents. More specifics about them soon.

We did meet you bodily in 1940, when you were taken into the small craft you thought a "cabin" at that time. Your companions were left frozen by hypnosis in the road and have no memory of the "full moon rising" at all.

You should be hypnotically regressed as soon as you can manage it, to dispel so many doubts in your mind and minds of others who are interested. Much will be revealed then to explain many other things. *Please try to do this soon.*

—6:30 P.M.

We can go into more detail later, but for now:

We have been monitoring you for many years, almost from birth. At age seven, Jamie and Amorto spoke to you about learning to write well and learning about people. It was truly the day of the snapshot of you under the lilac tree!

We observed you firsthand on many occasions after 1940, but never to your knowledge. By "we" I mean our colleagues—those who work closely with us but are of wholly Earth parentage. In this past year, you have been "followed" or "guarded," by someone close to you at all times. Because of our telepathic contacts we can change your "guards" many times a day. We have many who help in this. Our Earth organization is far greater than any of you realize, for we, by one means or another, always have persuaded our helpers to keep separate from each other. Therefore no one suspects the extent of our influence on Earth people.

But now the time has come to start putting it all together and the first move is to get our various contactees into close working order with each other....

Each of you separate people are doing a different type of work, and therefore have contact with different entities. Each of you is being developed to do a solemn task, each different, yet all synchronized. It will be some time before a clear picture emerges, but each day, each step, brings you closer to the time of complete comprehension.

As you have so clearly pointed out, Ida, we do not give you lines of preachment in order to teach or develop you. We make you live through events that do the teaching and developing—and what is specifically taught (or, rather, learned) depends specifically on what you [contactees] already have in you. Therefore we work with each of you in different modes and manners, to develop you to do tasks each different from the other, all possible to each of you in your separate being....

The reason you will find it hard to correlate your experiences with us (on the basis of events gone through, factors of contact, lessons learned, problems and pleasures derived from your experiences) is that there were altogether different types of each of these for each of you.

Each individual working with us does so in his or her own unique spheres of influence, action and outcome. This is so hard to put across to those who try to correlate various contacts and contactees. The correlation must take place on a different level of comprehension—not by studying contactees in too great of a depth, not by studying the factors of contact too strenuously, but by studying the *possibilities of use of that which is learned by the contactees.*

You are indeed beginning to accept some things as facts—the reality of UFOs, the humanness of your contactors, and, perhaps just a little, the good intentions of us all. When you get that skepticism licked, you will be a full-fledged worker for the Great Plan.

The plan has to do with the rejuvenation of the Earth and the further development (or evolution) of all humanity.

What you have learned working with us and how you have developed will be studied in great detail at a later date. You have too much to do otherwise right now.

For now, thank you and goodnight... *Your Hweig.*

February 21, 1980

Dear Leo:

Hweig has read along with me your proposed paper to be read at the MUFON Symposium in Houston on June 7th. He is particularly intrigued with the phrase "Captive Collaborators."

"You are exactly that," he says, "but only because you want to be. Each of you has, sometime in his or her life, asked to be of benefit to his fellow man, to be of some service to the world. So we give you the opportunity to be just that! Where else, how else, could you quiet, little, unobtrusive people find a way to be of so great a benefit to your people and to your world? Who else would give you the years of encouragement, the tedious years and years of guided development, the constant attendance and monitoring and direction of your intent?

"We offer opportunities no one else could possibly do, as we have great and mighty colleagues to call upon, personalities of many natures other than what you call 'human,' forces and powers you never dreamed of. We give you opportunity, direction and developed capacities to do just what you have prayed

and begged to do—help your own people, save your world from destruction and terror."
    Thus speaketh Hweig.

    Ida

Hypnotic Journey

The following is an excerpt from one of my letters to Leo Sprinkle:

[On February 25th] ...*Hweig and I got into a fight. I said that if I wrote what he wanted me to, it would sound like I had a head full of scrambled spaghetti instead of brains. He said I was too stubborn, so I quit typing and went downstairs. So there!*

The reason I include this is to show that we do have free will. We can always tell 'em to go to hell—and I sometimes do! Hweig never gets angry or scolds or is impatient with me, though he sighs a lot. And I am quite sure that sometimes they are holding a carrot in front of the donkey.

All of this leads me to the worry for today, this bit about retribution/bribe. Do they really play petty tricks on us if we do not collaborate properly? Personally, I think this is sheer superstition. Or maybe Hweig just treats me better than other contactors treat their contactees.

On the other hand, I have been having a streak of rare good luck, and it isn't because my brains have improved that much. Things just seem to fall into my pocket. I wanted a painting for a particular spot in my dining room and imagined just the sort of thing I could best use—Bill trots home from the flea market with a very nice picture that he bought for $25. It does not belong in the Louvre by any means, but is perfect for the vacant spot. I went to an estate sale in a nearby city and got there pretty late. So many people attended these sales that each is given a number as they come and go into the house according to their sequence. I was number 204, which was pretty dismal. But a girl standing next to me said she was tired of waiting and would sell her ticket for $20. It was number 30. So I got in among the first, and was thereby able to buy a set of sterling silver for $450. It is worth, by today's price in scrap alone, $1500. And I bought gold teeth for $100 that are worth $400. Never in my life have I had this kind of luck!

I tell Hweig I will not be bribed. I will do what I think is good and right, but I will not be bribed to do anything else. He says he has helped, and in one instance I know he did, but the help is given after I have done something to please him.

Relative to this, he says all the ufolk are in my debt because I once delivered to them a message concerning something they had been trying to discover for years. I have no idea who gave me the message. Hweig says I am to tell you about this now because something is going to come up in the near future and the information will be important to you. Oh, I hate this mysterious way of doing things! (This was about the secrets of Atlantis "off the Island of Crete," etc.) Hweig tells me that, with their instruments, they have determined [the records] are really there, and have been able to record and decipher them without moving the stones on which they are "written." "They will be found by man in time, and before long," he says.

He tells me he could not tell me what I was to photograph in Hawaii because they wanted to see my reaction to it first. And I was terrifically interested in, and quite excited by, the petroglyphs. I have always been very interested in such things, but never had the opportunity to view them directly. Hweig has been saying my job is to "interpret" them, the ufolk, to my people. Now he says this is to be done specifically by interpreting the petroglyphs, which were there long before the Polynesians came. Of course, I don't know what they say, but he does and, through me, will deliver the information.

I must say the Lord and Hweig move in mysterious ways! It was Bill who suggested we take Lee along with us this time and it was Lee who suggested we go take pictures of petroglyphs. Between us we probably took 140-150 pictures. We found some we could not find any mention of in all the museum booklets and other material we bought. Here again, Lee, on pure impulse, swung off the highway onto a side road that led down to a secluded beach. We picked up two bushels of fine coral and then went out on the lava rocks as the tide went out. We found some strange markings. These were not pictographs of people, fish, goats or sailboats, as we had found before, but were long, straight diagonal lines that ran into and intersected each other. We thought they indicated lines of force and pathways of energies we know nothing about. They will be important a little later, and we are to get at least five-by-seven enlargements of our "petro-pictures" to compare with others from elsewhere.

There is a lot more detail I could throw in here, but this is enough for today. It shows the direction in which I seem to be impelled—or shoved—at this time.

Ida

March 5, 1980

Dear Leo:
A comparatively short note to confirm plans for April and May.
...feel free to record or to have anyone else present you wish on the April 19th business.
My only fear is that this might be one of Hweig's jokes or teasing or wild goose chases he sometimes sends me on. He insists it is all true, real—both the 1940 meeting with a UFO and his intention to communicate vocally with you through me. He seems truly excited about attempting this. I do hope it is true and successful. As far as I am concerned, if anyone else is interested in any degree with such goings-on, they are quite welcome to be present. Set it all up any way you like. I am really shy, and do not talk much, but if my very talkative daughter is present, she will bolster my morale. If Lee cannot come, I will come alone. Bill is a somewhat negative influence. Lee is neutral, with a leaning toward the positive.
We are lost-in-the-crowd, nonsignificant, unpretentious people without much to recommend us except sincerity. We are neither young and beautiful, nor wise—simply terribly curious and persistent.

Ida

On April 5, 1980, I received a letter from another contactee. It contained a paragraph that had been delivered to her by the ufolk. They spoke of wanting to strengthen our race with their intelligence through mental telepathy and interbreeding of genes: "This has been achieved by mind transfer at birth of one of your children. Through this transfer, we then have contact with the occupant. When the child reaches an age for mental development, we then begin instructions." And there was more.

Just the day before, I had written another contactee who wanted to know if my hypnosis session with Dr. Leo was to be about UFO business only. I told her it was this time, but that someday I wanted to be regressed to my birth scene.

I was born the second of twins, a blue baby. I was pronounced dead at birth by a somewhat inebriated doctor. He wrapped me in a blanket and told my father to dispose of "it." After all else was tended to and the doctor gone, my maternal grandmother unwrapped the expired one (me) and gave me mouth to mouth resuscitation—and here I am. I had been wondering if I am really I, or some secondary attempt at life. Amorto told me once I was not a "Star People" but a "Something Else." Can that something else be a *mind transfer?* Is that why these fellows have asked me, over and over and over, "Ida, who are you?" Perhaps, I wrote in my journal that day, I should try to find the answer to this at the hypnosis session on April 19th.

On June 24th the communicators said, "We laugh at you for what you are! You are the *one* who now becomes the interpreter." And on July 29, 1979, Hweig mentioned a particular type of personality, come into the world to interact with the UFO people, adding strangely, "It all lies dormant in your mind."

Am I, then, a Mind Transfer?

HYPNOSIS SESSION OF APRIL 19, 1980
University of Wyoming, Laramie

In this presentation of the first hypnosis session of April 19, 1980, Dr. Sprinkle's hypnotic procedures have been minimally retained. Other material has been very slightly edited to delete repetitious, clumsy or obscure words. Other than these few minor changes, the transcript renders a portion of the recording precisely.

Dr. Sprinkle: I am Leo Sprinkle. I serve as director of Counseling and Testing at the University of Wyoming, and today we are filming a session with Mrs. Ida M. Kannenberg. Accompanying her is her daughter, Mrs. Lee Crawley. Also we have with us, as cameraman and all-around good man of many talents, Michael Lewis, who is filming. Our purpose here is to interview Ida Kannenberg with hypnotic techniques to see what we can determine about some early experiences in her life, and whether they have some connection with more recent events concerning information which she is gaining. So we'll turn to Ida and ask what information she would like to see if we can determine today. What things would be useful for us to explore?

Ida   Well, the first thing I would like to know: If the object we saw on the California desert in 1940 could have been a UFO? We thought at that time that it was a full moon rising, but we have checked the dates and the full moon arose a good ten days before we saw this object. And also, later that same night we saw a quarter moon in the same sky, and that doesn't quite add up.

Dr. S.   O.K. So one question is: "What happened on that December 1940 incident?" And another topic?

Ida   I've been told that when I was a little girl (about seven years old), I was playing under a lilac tree and two men approached. All that I can remember is that they asked me where a certain person lived in that neighborhood. I was able to tell them; however, they did not go to his house. Later they got into their car and drove off down the street. Now my present communicators tell me that, at the same time the two men told me (I supposed subconsciously, or in some fashion I did not hear), that I was to study people, to learn to write well, and that some day they would see me again.... I think they also told me to be a good girl.

Dr. S. That sounds appropriate, I suppose. O.K. Learn to write well and to study people and that they would meet you again someday, or talk to you?

Ida Yes.

Dr. S. So that is another incident that would be useful to explore in terms of memories with hypnotic procedures. Another topic?

Ida There is some question of the time when I was born. I was pronounced dead by the doctor and it was a good half hour later before someone took pains to revive me—and I just wonder how this could be brought out through hypnosis.

Dr. S. O.K. We'll follow along on that same sequence. Let's practice relaxation procedures and then when we're ready we will go back to that December '40 incident. *[Adjusts chair, then goes into hypnotic procedures.]*

Dr. S. ...feeling pleasant and comfortable now. Ready for the journey?

Ida Yes.

Dr. S. ...drifting back, back through time. And now you will be able to talk and describe your impressions. What are you experiencing now?

Ida Nothing.

Dr. S. Nothing happening right now?

[A portion of preliminaries is omitted here. It is not germane to this story. See *UFO's and the Psychic Factor*[1] for complete transcript.]

Ida Someone wants to speak through my voice!

Dr. S. O.K., fine. You can allow someone to speak through your voice.

Ida *[In a deep, heavy voice.]* War will not be tolerated. It endangers all, other planets, other peoples. Millions, millions of people. It will be stopped by psychic means, not weapons. Many are ready to help. They know not what

1. Kannenberg, Ida. *UFOs and the Psychic Factor.* Newberg, OR: Wild Flower Press, 1992.

Dr. S.

Ida

they are to do now. They will be told. That is all. Later. Now the other events may be told.

Dr. S.   All right. Thank you.

Ida   *[Pauses.]* O.K. We're ready.

Dr. S.   This time looks like we're ready to relax and go back to the experiences. Had to get the introduction first.

Ida   Somebody had to get his two bits worth in first.

Dr. S.   Right. When you're ready just lean back and relax. ...Back to the 1940 incident in December. Feel yourself be there vividly and accurately and b e able to describe your impressions and the events of that particular day and that particular night.... Feeling relaxed and comfortable now?

Ida   Yes.

Dr. S.   ...What feeling or thought is paramount about that incident.

Ida   Strange.

Dr. S.   Strange. Seems strange. O.K. Just focus on that feeling of strangeness.... What comes to mind when you focus on that feeling of something being strange?

Ida   Presence...presence of someone...can't review it visually. Can remember...some.

Dr. S.   Some. O.K.

Ida   Not forest fire. Too deep, too red, too even. Just even glow, red at first, very dark red, then the round shape comes from behind big rock, out to left. I must see it visually to continue....

Dr. S.   O.K. Be aware when it is appropriate, when you are ready you will be able to see it visually. Focusing on that strangeness, on that red spot...picture yourself viewing it. Let's see what happens next.

Ida   *[Sighs.]* Too much interference.

Dr. S.   Interference? Thoughts or feelings?

Ida   Presence.

Dr. S.   ...Just ask the presence if it is all right if you review these events. Ask for permission to review these events and see what happens then.

Ida   I say, "Are you gonna let me see this or not?"

Dr. S.   What's the verdict?

Ida	Not yet...something forgotten...who forgot? *[Pauses.]* They forgot.
Dr. S.	Something that has to be remembered first before we can be able to go back to this? Do they say what they forgot?
Ida	Subconscious closed...they use superconscious.
Dr. S.	Can they give permission to open up the subconscious?
Ida	No. Lets other entities in. Interference. Dangerous.
Dr. S.	But if we open up that information while protected, is that possible?
Ida	Not enough protection...superconscious can be used.
Dr. S.	O.K. Is permission given for superconscious to be used? To recall?
Ida	Yes.
Dr. S.	Let yourself use the superconscious level of awareness to recall the events as you're watching the round red spot. What happens next?
Ida	We stop side of road. Left side. Men go down road. Front of car. Headlights. Stay long, long, long, time. I think I'll rest. Doze. *[Pauses for a long time.]*
Dr. S.	What happens as the body dozes?
Ida	Who are you? *[Voice strong but quavery, as though held steady by effort.]*
Dr. S.	Are you speaking to me?
Ida	No, I speak to someone. Them. I'm frightened.
Dr. S.	You speak to someone and you're frightened.
Ida	Two. They need help. Ask me help...very disturbed themselves...come help. I can't visualize this. I only re-member.
Dr. S.	But you do remember the impressions. O.K. Let the im-pressions flow.
Ida	I ask what kind of help. Problem—what problem? Some-one hurt, needs blood transfusion. I say my type might not work. They say they know my type. Is O.K. It is a rare type. They are almost frantic themselves. I forget to be afraid. I go with them. Round cabin. I think cabin. *[Words lost while tape is changed.]* Inside light. Every-thing white metal. Sit on high stool. Left arm. Take blood. Big—syringe?
Dr. S.	Like syringe?

Ida   Yes. One—two—three—four—five men...look like us.
                One injured, bleeding. Chest. Shorter than others, cov-
                ered blanket. Can't see much. Blanket metal too!

Dr. S.   Blanket is metal?

Ida   They ask help other ways. Want to communicate again.
                Press something into my ears...way in, hurts. My nose,
                left nostril. Can't see. Way up, way, way up, hurts, not
                too bad. Can't see features. Can't remember. Taller than
                I. But look like us—human, like us. Only one speaks to
                me. Others look each other, nod, seem to talk—know
                without speaking. One speaks to me. English. No ac-
                cent. They tell me saved his life. Many thanks. They
                contact much later to help again. Vaguely see. Very,
                very vaguely. Nothing clear. More impressions than
                sight. Some sight. Take me back to car. Later I look at
                watch. Think I've been asleep forty minutes. Decide I
                looked at watch wrong first time, couldn't be that long.
                Fellows come back to car. We see quarter moon in sky,
                half way between quarter moon and new moon. So the
                big round red thing we saw before.... How could there be
                two moons? Doesn't make sense.

Dr. S.   O.K. Doing well. Just relax deeply...be aware that later
                on we'll have more information about these impressions
                as expressed...Now allow yourself to go back to the time
                during childhood...about your experience that may
                have occurred around seven years of age...of talking
                with, meeting with some strange men. See what impres-
                sions come to mind.

Ida   I am going down Carey Avenue. We live there, in Daven-
                port, Iowa. I am going home. There were no violets to
                pick. Only garter snakes. Maybe there's some lilacs left,
                but I don't like to go over where they are. But I will go. I
                will take my mother some lilacs, there weren't any vio-
                lets. They're gone. Too late. I don't like to be here.
                [Sighs.] But I think I will sit under the bush awhile. No,
                I'll lie down. I like to hear the ground. I think I'll go to
                sleep a little. But someone is coming. I hear footsteps in
                the grass. So I sit up. I didn't wear my glasses to play,
                so I can't see too well. Two men. One, big man, gray hair.
                One younger man, smaller. Business men, business
                suits. Probably rich men.

                They ask me, "Do you know Mr. Barton?"
                [Mumbles to herself.] Frederick Barton? Oh, Mr. Barton.
                I answer back, "Yes, the blonde man."
                "Doesn't he live around here somewhere?"
                "Right there. That house."

"This one right here?" The older man points to the house on the corner.

"Yes."

They talk together, I can't hear. Can't understand because I can't hear too clearly. Mumble.

The older man speaks again to me. (The younger man speaks only to him.) "What is your name? Where do you live? What does your father do?"

It is hard to remember. What is he saying?

"Where do you go to school? Do you like school?"

Why do they always ask that! He says something, but I don't hear him.

He's saying it like inside my head. "You will need to know all about people—why they do what they do, why they act as they do, what do they mean besides what they say. You will learn to write well; try harder at your spelling. Learn about grammar. Make nice sentences. Someday you will write many wonderful, unknown things. We will see you again when you are grown up and much older." Then he says out loud, "Be a good girl." They always say that.

They go away to their car. It's parked on the wrong side the street. They must not know they can't park on that side the street. Then they drive away. They don't go to Mr. Barton's house.

I start home and meet my mother coming. She has her camera. She says, "Who were the men talking to you?" I told her what I knew they said. She said, "I just wondered."

She said, "I have one picture left. I want to take it so I can send them to be finished."

So we went back to the lilac tree and she took my picture. I had on my boy scout uniform. ...They call it a play suit!

She said, "Supper isn't ready, but it will be in a little while. Come home pretty quick."

I don't have any flowers. But I go home anyway.... That's all I remember.

Dr. S. O.K. You're doing fine.... Let yourself relax deeply now, knowing that you will be able to go back to the birth experience and you may be able to talk and describe the impressions, even though you may not have been able to talk at that time.... Let yourself relax, deeply....

Ida *[Pauses for long time.]* Big room. Wide boards. Little rugs. Big bed. I am not in bed. I'm up there. High. I asked to come here. It is my duty. I must be born here. Not as the babies are born, but into the understanding

and consciousness of the one who is dead. I will be her—
I will be her mind, her consciousness, her feelings. No
one will ever know that I was not born. The body I occu-
py was born, but that was not I. They must revive that
body to breathe before I can enter. I wish they would
hurry up. It will be too late. Now. Now is the time. I must
forget all that went before. I must forget who I am. I
must not remember until it is time to remember—when
all the others will be called upon to know, for they are
many. It is so long waiting. No, not now. This is not the
time. I must wait. The time will come for all. Not now.
[Repeats loudly, with strong emotion.] Not now! *[Pauses,
then comes out of hypnosis.]* Hi!

Dr. S.	Hi! How you doing?
Ida	I'm not sure. I'm not sure.
Dr. S.	Experiencing some difficult feelings there?
Ida	No. Now that I'm awake, I'm not sure.
Dr. S.	Um, I see. I was wondering when you said, "Not now," whether you were talking about the way you felt in this situation or how you felt at the time of the experience?
Ida	Now is not the time to know.... I feel like I'm making it all up.
Dr. S.	*[Laughs.]*
Ida	I do. I really do. But I'm *not*.
Dr. S.	That's right.
Ida	I don't know that much.
Dr. S.	The feeling you had was like an out-of-body experience and looking down?
Ida	Waiting. Just waiting. Waiting somewhere. Somewhere in the room, waiting.
Dr. S.	To go into the body. Then the feeling was that you had a special mission or purpose or task?
Ida	Duty. Duty is not a mission. There was a choice. I could come here or stay. The other person chose first to stay and work from that side. I suppose I came here to work from this side.... Guess we gotta get together one of these days.... Bridge.
Dr. S.	A bridge. Between that life and that world and this world and this life?
Ida	To interpret. That is my duty—to interpret. I don't know what.

Another hypnosis session began with Hweig again speaking through my voice and answering questions for Dr. Sprinkle. This was also filmed and recorded, primarily for presentational purposes.

One of the questions dealt with the prevalent belief that many natural catastrophes, if not total destruction, were imminent.

"Not total destruction," said Hweig in essence. "But localized catastrophes. Some lives will be lost, but not millions. If people will listen to the warnings of your scientists, many lives can be saved."

This session took place on April 20, 1980. On May 18th, Mt. St. Helens in southwestern Washington state blew her top with devastating effects, and the foreboding of natural catastrophes became a reality.

Since I live not too far from the the volcano, I hear first-hand accounts and observations of the disaster. My pen grits upon paper as I write this. (It is impossible to keep the talcum-fine ash out of the house.) Our area received only half an inch (in wind-swept places) or an inch (in piled-up areas). Here a nuisance, elsewhere a total disaster. And St. Helens is not through by any means—and she is only the first in our part of the world. What is next—and where?

The reason I could not visualize when I expected to was because we were using the superconscious, which does not utilize the five senses! It is pure, direct mind knowledge, something like intuition, only stronger and impossible to fabricate. Because I could not visualize, I was worried that I might be making it all up! I knew I wasn't, though.

What follows is a paper I read on May 23, 1980, at the First Rocky Mountain Conference on UFO Investigation, at the University of Wyoming in Laramie.

In a hypnosis session with our Dr. Leo [Dr. Sprinkle was the director of the conference], I have been made aware that my contacts with UFO people have been a lifetime, indeed a before lifetime thing(though I was quite unaware of its extensiveness).

In many months of actual telepathic contact, I think I have run through every known emotion in response to them—from aversion, disgust, anger and fear, to total acceptance and even gratitude. For I feel the long term experiences have been for one overall purpose—to develop me as a useful "outer" person, and to give guidance to my evolutionary progress as an "inner" person. For that seems to lie behind the point of their contacts in our lives. They have said, "We come to rejuvenate the Earth and to aid in the evolutionary development of all humans."

This is the Great Plan with which they collaborate, but they have themselves many needs and wishes of their own—their own personal axes to grind. They ask bluntly that we exchange our help for theirs.

Since we are left exactly in the dark as to these intents and purposes (and of our before-lifetime dedication), we feel we have been invaded or imposed upon without justice or apparent reason (and sometimes without mercy).

Their methods to make contact were meant to emphasize their "realness," to make us aware they were actually there, that they have powers we have forgotten, and to make us, above all, *think*. Mostly they have succeeded in scaring us to death by their abrupt methods without explanations. My own first overt contact with them, in 1968, landed me in the state hospital, one scared-to-death kid!

Ten years later, in 1977, began the months of telepathic contact, writing as dictated by them, explaining many things. My skepticism stood firm, though I wrote avidly. I frequently referred to them as liars (not understanding how their various pieces of information fit together). To me, there was no fit, but it was only because I lacked enough of the right pieces to make them fit. After a hypnosis session with Dr. Leo in April, just past, my skepticism abated, so I am not kicking and squealing at every idea they try to impart.

I would say the events they put us through are meant to develop us through experience, the only way to true actual development. Each of us is taught what is uniquely "our thing." When enough contactees come forward with their complete stories (some of which they do not now remember, for they are timed to remember), then corroborations and coincidences will become apparent, and an overall pattern will emerge, which may be utilized to make sense to our kinds of minds. We will understand what they are trying to say with their kinds of minds. (Actually the kinds are the same, for both are human, but there are some light evolutionary differences, and some very large cultural and environmental differences.)

When all contactees come forward and share their experiences we can then see the underlying principles that hold it all together, and then it will become believable and understandable. We will see that we are not invaded—that we are but extended in experience and knowledge.

That is my personal conviction.

As I was flying home from the conference, I received the following charming, but baffling, poem from Hweig:

> We are not so far apart.
> I hear the beating of your heart.
> I hear the sighs and feel the tears
> And know the story of the years
> Has grown too long.
>
> Too far the morning of delight;
> Too long the heartbreak of the night;
> Too slow and tedious the flight;
> Too cruel the song.

Can we ever smile again,
Joyous through the years and pain,
Knowing that we are the twain,
The true, the strong?

Believe with me, we are the best,
Those who persevere the test,
Coming to the Ancient Crest,
Where we belong!

...Hweig.

June 10, 1980

Dear Leo:

Ever since the hypnosis session in April, I have been stewing and fussing. The description of myself on board the UFO is just too far-fetched for me. I mean the blood transfusion bit. It seemed false somehow to me at that time (or at least not real in entirety), and I have nagged at Hweig until he has confessed that the need for transfusion was only a play they used to get me on ship, and the scene of drawing blood was a hypnotic illusion to carry out the ploy. I just cannot remember having any physical marks or feelings afterward of anything physical having been done of that nature.

I did have a conscious recollection of something—somewhere going up a pebbly incline toward a "lighted cabin" and my left foot slipping backward on a pebble and someone barely touching my left elbow to steady me. Someone walked a pace behind, to my left, and someone two or three paces behind to my right. There may have been another, several paces in front, to my left, but always in a shadow (or perhaps in dark clothing), so I felt (or sensed), rather than saw his presence. A very tall, rigid figure stood by the door and did not move. I thought all this was a dream for a long time, until these characters told me I had actually entered a UFO (the lighted "cabin").

It was Amorto who touched me on the arm to steady me. Hweig was the injured person. He had tried to leave the craft but had stumbled, fallen and hit a sharp rock with his chest. He had the wind knocked out of him, but only a minor scratch and bruise where he hit on the rock. They used this as a method of getting me into the craft. They were handicapped by instructions not to force me physically and not to use any of their psychic methods of getting me aboard. I had to go of my own free will.

They had been studying my dossier for a long time, learning my ways, manners, speech, etc. But they did not know who I was (or when or where they were to contact me) until practically the last minute. They knew me well enough then to know that the only way I would accompany strange people in the middle of the night under such circumstances was to make me think I was drastically needed to help someone, and that there was not an instant to spare.

Hweig credits Amorto with the actual idea of saying someone was hurt and needed a transfusion. That certainly would be about the only thing that would make me go willingly with them.

Now this begins to make sense, and while it still does not seem really real to me, it does seem more plausible—that is, as something I can understand myself doing.

There was no blood taking or transfusion—only an impressed illusion of this to carry out the ploy. Their purpose in getting me aboard was the implantation of the devices as stated. These can be detected by X-ray, they say (but do not advise a nose X-ray, as it is too close to my eyes, which are very poorly sighted to begin with). Ear X-rays should reveal very tiny paper-thin "wafers," oval and about an eighth of an inch long or a bit smaller. Sometime I hope to have the opportunity to check this out. You see, I am still retaining a cautious attitude, in general. Hweig is not at all adverse to fabrications in order to check me out on something. This angers me exceedingly. "I tell you guys only the truth; why can't you be equally truthful to me?" [I tell them.] And, of course, he says it is because time is so limited, and they have to "take shortcuts across the truth," and sooner or later, "when the time is right," I will always be set straight on things.

"But what if I have passed on lies to other people in the meantime?" I ask. "I do not intend to be a liar for any reason or purpose!"

"We know that," says he. "What we allow you to tell others is always, in essence, truth."

Please note word "allow." And damn their "in essence"! I get so mad sometimes!

What disturbs me most about the event on the UFO (if such an event there was—and I am willing to agree that there probably was) is that they won't let me see it visually. I think that bit about "to free the sub-conscious is dangerous" is another bit of hokum. The real reason is they don't want me to know or remember some things yet. Why can't they just say so?

There now, I've gone and worked myself up into a real mad!

Sincerely,

Ida

Footprints in Time

On September 20, 1980, I wrote a letter to Colonel Wendelle Stevens that included the following:

> *...I am told by my monitor, Hweig, that I can now be hypnotized and recall visually what I could not see the first time. I knew there must be some powerful reason why they did not want me to see the entire episode within the craft, and I thought and thought and thought. Of course, dumb Ida, there was another implant—a frontal lobe implant. They were not ready for me to know this. Now they corroborate my guess.*
>
> *The frontal lobe was no doubt reached through the eye socket, as our surgeons do a lobotomy. No scars. Hweig tells me all implants should show up on very careful X-rays. The doctor will think I am crazy (but I sometimes think that myself).*
>
> *...The brain implant permits them to control the entire molecular and muscular activity of my body. This explains how they are able to make me laugh, the constant fine vibration I feel throughout my body, the movement inside my forehead, and a few dozen other things.... So much is explained by such an implant—why they fly up the wall every time I sneeze or whistle, how they help me carry a tune, and a great number of other things....*
>
> *Ida*

September 24, 1980

Dear Leo:

> *Yesterday I went to the doctor for skull X-rays (one frontal and one left side). Hweig has been bugging me for a long time to do this, and after Col. Stevens asked me if I had done so, Hweig said I must go at once! So I did.*

The doctor did not look at me as though I were crazy at all. He said his wife read everything she could get her hands on about UFOs, though he didn't pay too much attention.

When he examined the film he said, "Clear as a bell," and "Nothing shows up, but it depends what the implants were made of. Plastics, glass or transparent material would bot be picked up by X-rays."

I had to tell him I did not know what the material was. In the background, Hweig was saying, "But they're there! We never would have urged you to have X-rays if they were not. It would only make you lose confidence in us." He sounded truly stunned and consternated (his pet word).

I said, "Well, what are they made of?"

"I always assumed they were of the same material that our UFO craft are made of," he said, as I went to lunch. "A metal as yet undiscovered on Earth, which we treat—the only way I can describe it is we plasticize it. It is supremely strong and wonderfully light in weight."

Hweig has been talking about the plasticized metal of their craft to both my friend and me ever since we left the conference. "No," he said, "I do not mean an amalgamation nor lamination of plastic and metal. I mean we take the metal and change the molecular structure. Oh, my god!" he says then. "Of course, changing the molecular structure could make it invisible to X-rays!"

"What does it look like?" I asked.

"Like translucent aluminum foil. The craft material is thicker of course, but we can make it translucent—indeed transparent—when we want to by lighting it in certain ways." (He had told my friend and me this before also.)

"Perhaps the light or rays of the X-ray made the implant material transparent also," I suggested.

"Let me ask our technical people about the material," he said.

So about five o'clock this morning he woke me up (the skunk) with this information:

"It is the same metal as the craft are made of, alloyed with another metal unknown to Earth at present. The alloyed material is changed in molecular structure to make the paper-thin, circuit-stamped wafer which we implant. It is translucent not transparent, but it is quite possible it could have reacted to the X-ray, as our craft material reacts to our lighting to become transparent. We are going to research this exhaustively."

October 3, 1980—11:28 A.M.

So many questions, my dear. Some I can answer now and some not yet.

First you ask about myself. I cherish your concern for me! I did say that as a small boy I was taken by the ufolk after my mother's death to be reared as one of them. My father was an Atlantean time traveler (not from the space stations, for I am far older than the time these appeared in your heavens). He was a time researcher from Atlantis. I know no more of him than that. He did not come through on the space stations....

Yes, I did say I have never been on Planet X until quite recently. So, obviously, as you observe, there must be some other place where the ufolk have established colonies and now reside. There are several experimental stations in your own planetary system, mostly underground. (We have explained this previously.) And when one of the ufolk says he is from a planet of your system, he is simply saying that was his last port of call!

The colonies are not few, but their inhabitants are *not* of Earth stock. These are some of the strange personages your contactees have encountered. The few Earth type persons are on Planet X. No one yet has been told the exact location of this, nor will anyone be told for some time to come.

Long ago I gave you a hint (knowing you had no way of checking it out). I will repeat it here: "Look for the sun that has six planets. All but one of these is in perfect geometrical alignment with its sun and each of its fellows. The sixth planet—Planet X, the Rogue Planet—has no seeming inter-balance with any of the others. ...No, it is not in your solar system."

My place of being, as I grew and was educated, was in a system of stars quite observable in your telescopes. There is a certain group of these stars that has one "lost" member. It is lost because of devices, which render it invisible to your scopes, and also give it other qualities you would not comprehend. This "lost" star hangs in full view, if you but had the right instruments to see. It is the one where I lived (so many years that you would reject the figure as a misrepresentation). Other contactees will have more to tell of this group of stars. Later I was sent to a space station to live among my own kind, the Atlanteans. I became a leader.

Second question: "How did the Lemurians, so long ago, get from Earth to Planet X if their space and time research was in its infancy?" Through the intervention and aid of the Creative Forces. They were taken in "thought ships," i.e., UFOs called forth by mind power. No corporeal body could be contained in, or traveled in, any of our (Atlantean) thought ships, but those of the Creative Forces are unbelievably superior. They can be made to do whatever the Creative Forces want done.

Third question: "How many Atlantean space stations are there?" Originally eight, now six. There were some experiments that turned out badly.

Fourth: "What became of the others?" They were lost. One was submerged in the Arctic. One was partially submerged. It has been glimpsed by pilots. No more on this now, please!

Fifth: "How large are the space stations?" From half a mile to many miles long. I cannot be more explicit now.

Sixth: "What happened to the five Avenger craft in 1945, the Mariner, and the Super Constellation in October, 1954?" They were kidnapped by crews from Planet X. Their personnel are almost all alive, adjusting well, and learning. They will be returned when it is safe for us to do so.

Seventh: "Why don't the occupants of UFOs tell us things straight out when they meet us?" We are hard put to find proper means and methods of putting our material across. We must utilize symbol and allegory to fit each individual mind. Did not Jesus speak in parables?...*Hweig*

October 25, 1980

Dear Leo,

Today we got a letter from my sister. A few days ago, she was out in the hills with her grandson-in-law and his uncle. The two men took their rifles and went hunting for deer. She stayed near the truck. She felt impelled to turn suddenly and look up over the hill—and there was a spectacular UFO! She said it had a greenish glow, surrounding it like a halo, which changed to a softer yellow. In the center of the glow was a silver, oval object. It "stood" there for just a minute or so, long enough for a real good look, then shot upward out of sight.

At Hweig's insistence, I have bought a pair of red boots to wear with my new Levi suit. "Don't you know what they say about red shoes?" I asked Hweig. "They say red shoes can take you anywhere. You must be very careful they don't lead you astray!"

"Exactly!" he says cheerfully.

Ida

In February of 1981, Bill, Lee and I again went to Hawaii. Bill toddled off on his own itinerary, while Lee and I drove, in our rented Mercury, to the beach where we had found the strange markings on the rocks. [See Photo 1 on page 204.] They were so different from the usual Polynesian petroglyphs. We took more photos, and tried at the museums (and again after returning home) to find someone to interpret them.

Lee kept insisting that they looked like ogham, a script originating (apparently) in Ireland, or at least one spread abroad by Irish seamen, and taken up and used by Arab sailors as well. At her insistence, I eventually contacted an epigrapher. On March 25, 1983, I received information that they were indeed ogham, and in the Hawaiian language. A star map of the Pleiades was identified, and some statements about the summer solstice were interpreted. [See Photo 2 on page 205.]

Sometime later, another photographer sent almost identical pictures to the same epigrapher, who now disclaimed that they were ogham, but said they were gas vents and other natural phenomena.

I observed that if the lines were made by nature, they would not be so carefully spaced. Instead, they would run into and crisscross each other and run off the edges of the rock. Two adjacent rocks split apart by time and weather would carry continuous lines. They do not. Margins are carefully observed and abrasions strategically placed. The lava was broken and weathered eons before the ogham was abraded (probably only a few hundred years at the very most).

I showed some of the photos to a handsome young Hawaiian who was installing a telephone. When he saw the double basins, he cried, "Kapu! Kapu! That is for the Kahuna. You do not touch those!"

The first time we had visited the beach, in 1980, Lee had thought she detected the remnants of a heiau (a Hawaiian temple) at the far end of the beach. But in 1981, the rocks had been tumbled about, and she said the outline was no longer there. We learned that one of the most influential Kahunas of the Islands had lived within half a mile. Putting all of our facts together, we decided the area was sacred to the Hawaiians. Therefore, I will not divulge the exact location (but I will do so to serious, conscientious and responsible researchers).

Hweig had nagged me incessantly to get good photos of that particular location. I studied the pictures over and over—and finally saw the reason for his interest.

"Is that a *footprint?*" I asked. "It can't be. The instep is lower than the heel and toe. In a real impression the instep would be higher." [See Photo 3 on page 206.]

Hweig said, "It is a historical marker. That is where the foot of true humans first trod upon Earth. Hawaii is a remnant of Lemuria. This is where "the sons of gods" first came to Earth. These were the first "true men.""

I asked, "Those who mated with the daughters of men?" (Genesis 6:4)

"Eventually this happened. Not necessarily at once."

"I'm befuddled," I complained.

"So is history. Don't worry about it. There are still writings to be found that will explain the whole process."

"Were the sons of gods physical men?"

"Of course. How else could they have mated with the daughters of the Earth?"

"How could they, genetically?"

"They could alter the genetic code. Even now, your own scientists are working on genetic alteration."

"Where were they from?"

Photo 1
Polynesian petroglyphs found by Ida and her daughter, Lee, in Hawaii in
February of 1981.

Diagram 1
nana Mola hoku

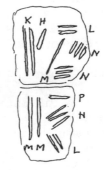

imo imo Lani po

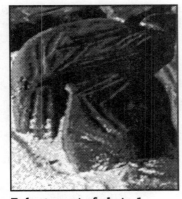

Enlargement of photo 1

Imo imo lani po—
"When the night sky is scintillating"

Nana mola hoku—
"Observe the revolution of the stars."

The ogham is read from right to left. Vowels
are omitted. (Hawaiian language)

Diagram 2
Star Map of the Pleiades

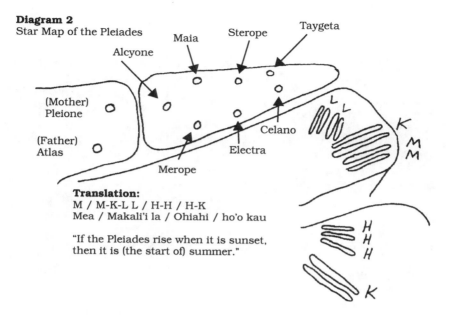

Alcyone — Maia — Sterope — Taygeta

(Mother) Pleione

(Father) Atlas

Celano

Electra

Merope

Translation:
M / M-K-L L / H-H / H-K
Mea / Makali'i la / Ohiahi / ho'o kau

"If the Pleiades rise when it is sunset,
then it is (the start of) summer."

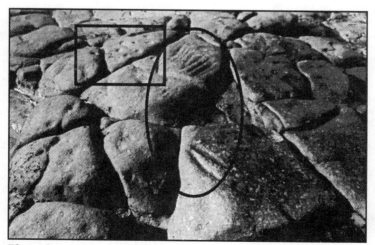

Photo 2
The photograph of the above diagram. Notice the small holes in the
square, as well as the straight marks outlined by the circle. Ancient
communications from above?

Photo 3
Notice the footprint in the black rectangle and
the marking outlined by the circle.

Diagram 3A
"The Hawaiian version." The
mysterious writings found in
the above photograph taken by
Ida in Hawaii.

Diagram 3B
"The original glyph." Symbol
as hand-copied from James
Churchward's book,
The Children of Mu.

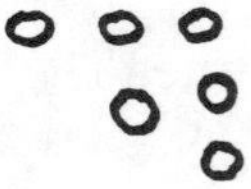

Reason Tao created one,
One became two,
Two produced three—
From these all mankind
descended.
(Tao Te Ching)

The footprint and "writing" have no connection with the ogham which was
put nearby much later. I believe that Hawaiians used this as a sacred
place because of the footprints.—Ida

"This is where the foot of *true man* first stepped upon Earth," said Hweig.
(NOTE: True man wore *shoes!*)

"Other planets. Other solar systems."

"Planet X?"

"Yes. Some of them."

"Is that why the UFO people of Planet X and their allies treat us as they do—as though they own us and can do with us as they will?"

"They feel they do own us."

"Like hell they do! I'm tired of men telling me they own me. God owns me!"

"Spiritually, yes. Genetically, no."

"Then the whole battle is over who owns the Earth human race—God, Planet X or the men of Earth themselves?"

"I'm afraid so."

But that was not all the photographs revealed.

Geological maps show the beach area having a fringe of rock from boiling mud—no doubt where the lava slipped into the sea at that point. The ogham area, which is adjacent, is on the basaltic lava. The "print," or marker, and the gas vents are on a rock with a slippery yellow surface.

On a section of the rock alongside the print is a series of indentations that look like poke holes. The formal arrangement indicates that they are not gas vents.

This careful arrangement of dots called to mind a page from *The Children of Mu,* by James Churchward.[1] The symbol given was very similar to the markings I had found on the rocks in Hawaii. [See diagram 3A and 3B, page 226.] This has been called the "mysterious writing." It can be read in three directions. Churchward identifies the glyph as appearing in many places throughout the world, but claims it originated in Mu (which we have called Lemuria).

Another source is Lao Tse in *Tao Te Ching,* a Chinese book written in 600 B.C.:

Reason Tao created one,
One became two,
Two produced three—
From these three all mankind descended.

Now I pinned Hweig down again. I said, "Relative to the mysterious writing and the seed of humanity, I have some questions. Will you answer them?"

"As far as I am permitted."

"Whether this is a footprint (and of the first true man) or not, and whether this is an example of the mysterious writing or not, my questions would be the same. Were the daughters of Earth the evolving creatures that anthropologists call *Homo Erectus?*"

"Along that order, but more advanced."

"Was the result of that union what we call Neanderthal man?"

"The forerunner. Neanderthal was a descendant long afterward."

1. Churchward, James. *The Children of Mu.* New York, NY: Washburn Publishing, 1931.

"Did the breeding with Earth beings happen only once?"

"Oh my, no. Over a long, long period of time. Many times."

"Were these first true men androgynous?"

"They were."

"Does the mysterious writing mean that one androgyne divided to become two—man and woman—like Eve derived from Adam?"

"Yes, it does."

"Did their first matings create Cro-Magnon?" (The two produced three).

"The forerunner of Cro-Magnon."

"Well, *Homo sapiens, sapiens* then."

"Precisely."

"The mating with Earth 'women' eventually led to Neanderthal, but mating between their own kind led to Cro-Magnon or *Homo sapiens sapiens.*"

"That is correct. But you are painting yourself into a corner. Let me extricate you."

"Please!" I could see I was getting confused.

"Gradually the two strains began to pull apart. As more and more of the purer strain was born, there was no longer a need for the choosing of mates among the original Earth born. The strain that was to become Neanderthal went along its own way as far as it was able to evolve, then died out. The purer, less intermixed strain chose, quite naturally, the more intelligent and physically appealing mates—and Cro-Magnon appeared, for a time, alongside Neanderthal. All persons now living are of some mixture. There were those who tried to maintain the purity of the true men but that is now lost. No one on Earth (or on Planet X either), is of the original purity. Until the reality of Lemuria and Atlantis are proven, there is not much use in telling you more." Hweig indicated this conversation was over.

I kept thinking of many questions. Had there been only the footprint, or only the mysterious writing, I would have paid little attention. But to find them adjacent to each other and then hear Hweig's interpretation was too much for me, in all conscience, to refrain from mentioning. Could the smudges on the edge of the rock, where the mysterious writing was not in perfect accord with the glyph, indicate the two Earth human strains that were to follow, or that did follow the advent of the true men? Was it put there while Lemuria was still part of Gwandanaland, or eons later, after Hawaii became an island? Hweig called it a historical marker and its proximity to the Hawaiian ogham would indicate it might be more recent. No matter when it was put there or by whom, the interpretations remain the same.

I have lain awake for hours trying to resolve a balance within my conscience, to omit any mention of the ogham, the interpretations, the pictures, or of the beach—to be true to my promise (to Hweig & Co.) to inform the people of Earth seeking a fair and equitable balance. Additional facts are on file.

I must leave all the questions where they hang—in midair. May they inspire curiosity and study. My personal beliefs are not important. What has been important to me has been to pass along what I have been shown and told to the honest best of my abilities, trying at all times to protect others from any potential embarrassment or harassment. Most of the names of family and friends are inventions. My story is not.

Epilogue

So what has happened during the nearly fourteen years since the foregoing was written? Has Hweig marched up, rung my doorbell and invited himself to supper? Hardly—although he knows that he would be welcomed.

Frankly, I wonder if Hweig's contingent will ever reveal themselves to us more openly than they already have. They prefer to work from outside our limitations. They dart in from time to time like celestial mosquitoes, stinging us to thought and action.

To engage more openly with us would entrap them in our concept of mechanistic time, embroil them in our political arguments and world-wide economic problems, and mire them down in our crippling belief systems. They are too canny to become involved in all that. To become too deeply enmeshed in our daily activities would cost them their freedom and power of action.

I suspect that someday, having scattered their seed far and wide over the Earth (I do not mean in a biological sense), they will sail off into the sunset without a word of farewell. We will wake up some morning in the not so distant future to find them all gone, and we will ask in bewilderment, "What happened?"

I wonder if they are not as weary as we are of their "catch me if you can" tactics? I will not move myself one inch toward the window if someone says, "There's a UFO outside!" I will say, however, "Bring it in the house and I will look at it."

For fourteen years, Hweig has upheld his role of a "live-in" companion—always ready to advice and encourage (or to tease). Others chime in from time to time—the Hidden One and some of the Interdimensionals who call themselves the Creative Ones, or the Shining Ones. I cannot identify them other than by these designations. The Hidden One has been presented to me so differently at so many different times that I cannot decide just who he might be. He did complain that others had used his name in the past and had caused great confusion for everyone involved.

I would like to speculate on the most bizarre scenarios some experiencers are claiming these days in all sincerity and sober belief. Can these possibly be scenes of "virtual reality" imposed on a person or even a group of persons by certain groups of ETs?

As Hweig pointed out: It makes no difference if the person had an experience or only thought they had an experience. The psychological effect is exactly the same.

Because the ufolk are so adept, so masterful, at imposing scenarios of virtual reality on us, I cannot take things experienced under hypnosis as trustworthy of fact. I hope I do not sound skeptical to the point of paranoia. I find this a very logical universe, and I cannot believe aliens from another material planet can play hop scotch with the universal rules of law and order. But they certainly can—and certainly do—play games with our minds and make us see, even feel, exactly what they want us to. I find this a chilling thought, and it makes me a little angry and never absolutely positive about anything.

Project Earth was actually my first book to be finished on the topic of aliens, but was held back because it was believed that *UFOs and Psychic Factor* and *The Alien Book of Truth* would be of more immediate help to people. Just today I received another letter saying, "Your books saved my life and that of my daughter."

UFOs and the Psychic Factor was finished in 1986 as a home-produced handbook, then was added to and published by Wild Flower Press in 1992. *The Alien Book of Truth*, the product of many years of study, was wrestled to a conclusion and published in 1993, also by Wild Flower Press. *Project Earth* was then cultivated from over a thousand typewritten pages to the final form which you see.

The reader most certainly still has dangling questions and open issues even after reading all three of my books. I have searched and puzzled and pondered these same questions for more years than I would ever admit to, and I still don't have all the answers. I am only a little less perplexed than I was many years ago. I have given up hoping that all the pieces will fit nicely together into one big jigsaw puzzle. Every new piece that fits dislodges another older one. Hweig has often said that no one gets the whole truth. Each of us must establish his or her own conclusions after studying many different sources. Each contactee seems to receive a piece of the puzzle, therefore, it is up to the reader to separate the wheat from the chaff. I wish you strength and stamina in blazing your own path.

Hweig asks me to thank all those persons who have enquired about him. He hopes the day will come when he can answer in person.

From both of us—
Go with God until we meet again,

Ida and Hwieg
January 31, 1995

The Planet Speaks

(a non-poem)

Does she know a cold, sick fear—old Mother Earth—
When her children blast her very womb?
Who heeds the shuddering pain,
The choking question, "Why?"
Her cries are soft. The quiet she cuddles to herself.
Where has she failed? What fault was hers?

She counts again her brood.
"Let's see, there's shining ant.
He's not the biggest, but oh! so dear!
This crisp blue flower reflects the sky.
There's Everest, worth anyone's respect.
The sturdy elephant, and Kiluala the impetuous one.
She means no harm, although unsettled."

Her count goes on, a rosary of love.
The wild, the docile, the rebel,
Each is equal in a mother's heart.
She comes at last to humankind.
She adds, subtracts, takes percentages,
Multiplies, divides by billions.
The equation will not balance.

"What lack is mine?
Why have humans turned away?
Have I grown so shabby, unpresentable,
They blush to acknowledge their old gray mother?
To be sure I've lost the purity of new.
There are frightful rents in the fabric of my being
Where minerals, gold and gemstones lay.
I'm scarred and bruised. My hills are bald
Where once great forests grew.
My veins run turgid with the sludge of industry.
The air I breathe, as do they, is thick and foul with fumes.
Where have I been negligent?
What have I reserved for self?
If only once I could hear them say,
'It is time we take responsibility.
Old gray mother, we love you.'"

So speaks the planet.

—Ida M. Kannenberg

Bibliography

Churchward, James. *The Children of Mu.* New York, NY: Washburn Publishing, 1931.

Fowler, Raymond E. *The Andreasson Affair.* Newberg, OR: Wild Flower Press, 1994 Reprint.

Kannenberg, Ida M. *The Alien Book of Truth.* Newberg, OR: Wild Flower Press, 1993.

_____. *UFOs and the Psychic Factor.* Newberg, OR: Wild Flower Press, 1992.

Roberts, Jane. *Seth Speaks.* Englewood Cliffs, NJ: Prentice-Hall, 1972.

Smith, Warren. *The Book of Encounters.* New York, NY: Kensington Publications Co., 1976.

Vallée, Jacques. *The Invisible College.* New York, NY: E. P. Dutton & Co., 1975.

Vaughan, Alan. *Incredible Coincidence.* Philadelphia, PA: Lippincott Co., 1979.

White, Stewart Edward. *The Betty Book.* New York, NY: E. P. Dutton & Co., 1937.

_____. *Unobstructed Universe.* New York, NY: E. P. Dutton & Co., 1940.